# First-Order Logic

## Raymond M. Smullyan

*City University of New York
and Indiana University*

Dover Publications, Inc.
New York

The research of this study was sponsored
by the Information Research Division, Air Force Office
of Scientific Research, under Grant No. 433–65

## Copyright

Published in Canada by General Publishing Company, Ltd., 30 Lesmill
Road, Don Mills, Toronto, Ontario.
Published in the United Kingdom by Constable and Company, Ltd., 3
The Lanchesters, 162–164 Fulham Palace Road, London W6 9ER.

## Bibliographical Note

This Dover edition, first published in 1995, is an unabridged, corrected
republication of the work first published by Springer-Verlag, New York,
1968 ("Band 43" of the series "Ergebnisse der Mathematik und ihrer
Grenzgebiete"). The author has written a new Preface for this edition, as
well as providing a number of corrections.

## Library of Congress Cataloging-in-Publication Data

Smullyan, Raymond M.
    First-order logic / Raymond M. Smullyan.
        p.    cm.
    "An unabridged, corrected republication of the work first published
by Springer-Verlag, New York, 1968 ('Band 43' of the series 'Ergebnisse
der Mathematik und ihrer Grenzgebiete')"—T.p. verso.
    Includes bibliographical references (p.    –   ) and index.
    ISBN 0-486-68370-2 (pbk.)
    1. First-order logic.    I. Title.
QA9.S57    1994
511.3—dc20                                                          94-39736
                                                                        CIP

Manufactured in the United States of America
Dover Publications, Inc., 31 East 2nd Street, Mineola, N.Y. 11501

*To Blanche*

# Preface to the Dover Edition

Virtually all the improvements in the edition are due to Dr. Perry Smith, who made many corrections to the earlier edition and provided some very valuable suggestions. He also called my attention to the fact that the first published paper on analytic tableaux was that of Lis [1].

*Elka Park, N.Y.*                                                                    *R.M.S.*
*September 1993*

# Preface to the First Edition

Except for this preface, this study is completely self-contained. It is intended to serve both as an introduction to Quantification Theory and as an exposition of new results and techniques in "analytic" or "cut-free" methods. We use the term "analytic" to apply to any proof procedure which obeys the subformula principle (we think of such a procedure as "analysing" the formula into its successive components). Gentzen cut-free systems are perhaps the best known example of analytic proof procedures. Natural deduction systems, though not usually analytic, can be made so (as we demonstrated in [3]). In this study, we emphasize the tableau point of view, since we are struck by its simplicity and mathematical elegance.

Chapter I is completely introductory. We begin with preliminary material on trees (necessary for the tableau method), and then treat the basic syntactic and semantic fundamentals of propositional logic. We use the term "Boolean valuation" to mean any assignment of truth values to all formulas which satisfies the usual truth-table conditions for the logical connectives. Given an assignment of truth-values to all propositional *variables*, the truth-values of all other formulas under this assignment are usually defined by an *inductive* procedure. We indicate in Chapter I how this inductive definition can be made explicit—to this end we find useful the notion of a *formation tree* (which we discuss earlier).

In Chapter II we give a detailed presentation of our version of the tableau method for propositional logic. Our tableaux—which we term *analytic tableaux*—combine certain features of the semantic tableaux of Beth with the tableaux of Hintikka. Our tableaux, unlike those of Beth, use only one tree instead of two. Hintikka's tableau method also uses only one tree, but each point of the tree is a finite set of formulas, whereas in ours, each point consists of a single formula. The resulting combination has many advantages—indeed we venture to say that if this combination had been hit on earlier, the tableau method would by now have achieved the popularity it so richly deserves.

Chapter III is devoted exclusively to the Compactness Theorem for propositional logic—namely that a denumerable set of formulas is (simultaneously) satisfiable providing all its finite subsets are satisfiable. We discuss several different proofs because they are analogues of different completeness proofs for First-Order Logic.

Chapter IV consists of purely introductory material for First-Order Logic. Chapter V treats tableaux for First-Order Logic. There we give proofs of the well known completeness theorem, compactness theorem (for First-Order Logic) and the Skolem-Löwenheim theorem.

Chapter VI is devoted to the unifying principle which we sketched in [2]. Some of the applications of the principle are treated in subsequent chapters.

Chapter VII is devoted to one theorem which should be far more widely known and appreciated than it appears to be. We refer to this as the *Fundamental Theorem of Quantification Theory*. It easily yields the Completeness Theorem and far more! It is difficult to credit this theorem to any one author. It is Herbrand-like in character, but it also incorporates many ideas due to Henkin, Hasenjaeger and Beth.

In Chapter VIII we show how our earlier completeness results can be used to etablish the completeness of the more usual Hilbert-type axiom systems for First-Order Logic. We start with a system $Q_1$ which is much like a standard system, though it incorporates some new features—viz. avoidance of worry about collision of quantifiers. We then consider several other systems, each of which is ostensibly (but not really) weaker than the preceding, ending with a system $Q_2^*$ of particular interest. The rules of $Q_2^*$ are all 1-premise rules, and it is not immediately obvious that the set of theorems of $Q_2^*$ is closed under modus ponens. The completeness of $Q_2^*$ might come as somewhat of a shock to those not already familiar with the Fundamental Theorem, but the completeness of $Q_2^*$ is really just about tantamount to the Fundamental Theorem. The system $Q_2^*$, though a Hilbert-type system, has features reminiscent of Gentzen's Extended Hauptsatz. It obeys something rather close to the subformula principle, and the only way a sentence can possibly be proved in this system is by starting with a *tautology* and then applying purely quantificational rules. A modification of $Q_2^*$ is given in Chapter XVI, which yields a new proof of the Craig Interpolation Lemma, reducing it to the propositional case without appeal to prenex normal form.

In Chapter IX we consider a modernized version of the Henkin-Hasenjaeger completeness proof, using an idea due to Beth, with a touch of pepper and salt thrown in by the author. It proceeds in a slightly more direct manner than any other version we have seen, as does also the proof there of the Skolem-Löwenheim theorem.

The material of Chapter X will, we hope, prove of expository value even to the experts. There we try to gain some insight into the essential differences and similarities between completeness proofs along the Lindenbaum-Henkin lines and completeness proofs of cut-free systems. We conclude the chapter with a new completeness proof, incorporating an idea suggested verbally by Henkin to the author, and modified so as to be applicable to cut-free systems.

In Chapter XI we study Gentzen systems for propositional logic and quantification theory. Using our unified "$\alpha, \beta, \gamma, \delta$" notation, we are able to formulate these systems in a *uniform manner*—i. e. in such a way that the logical connectives and quantifiers do not appear explicitly in any of the postulates. This allows a particularly attractive treatment of the metatheory, since 12 cases can then be collapsed into 4.

Chapter XII consists of Gentzen's Hauptsatz and related results. Our unified notation pays off further in simplifying the proofs.

Chapter XIII treats the tableau method for *prenex* formulas. For prenex formulas, the tableaux need no branching! The resulting completeness proof is substantially that of Dreben-Quine (cf. appendix to Quine [1])—or of Patton [1]. This is very closely related to Gentzen's Extended Hauptsatz (as we see in the next chapter). In concluding Chapter XIII, we sketch an interesting alternative completeness proof which avoids appeal to any *systematic* construction of the tree, but rather utilizes the Henkin-Hasenjaeger principle in a somewhat different context. [In general, we strive throughout this study to consider various inter-relationships among the ideas of many authors.]

Chapter XIV contains new material on Gentzen systems. After first relating Gentzen's Extended Hauptsatz to the Dreben-Quine completeness proof, we then consider a new version of the Extended Hauptsatz which does not require appeal to prenex normal form. Then we consider a Gentzen-type system which satisfies a principle stronger than the usual subformula principle—i. e. the system is such that if any sequent $U_2 \rightarrow V_2$ is used in a proof of a sequent $U \rightarrow V$, then every term of $U_2$ is a subformula of some term of $U$ and every term of $V_2$ is a subformula of some term of $V$. This feature is crucial for applications in the remaining three chapters.

In Chapter XV we use this new Gentzen-type system to obtain a very easy proof of the Craig Interpolation Lemma. For the benefit of the non-experts, this chapter concludes with the usual derivation of the Beth definability theorem as a consequence of Craig's lemma.

In Chapter XVI we consider new and stronger versions of the completeness theorem which we call "symmetric completeness theorems". These occured to the author as a consequence not of Craig's Interpolation Lemma itself, but rather of certain ideas used in some proofs of the Interpolation Lemma.

Our final chapter contains new systems of Linear Reasoning. We consider 3 such systems—each of which is related to one of the 3 "symmetric" completeness theorems of the preceding chapter. The first system (which we treat in greater detail than the others) does not require appeal to prenex normal form, and uses the idea of "configurations" introduced by the author in [2]. Our second system comes somewhat closer to the original system of Craig (and does appeal to prenex normal form). The third system uses neither prenex normal form nor configurations, and is closely related to the Fundamental Theorem proved in the preceding chapter.

The author wishes to express his warmest thanks to Sue Ann Walker, Robert Cowen, Melvin Fitting and Edwin and Steven Rosenberg for much valuable help in the preparation of this manuscript.

# Contents

## Part III.  Further Topics in First-Order Logic

Part I

**Propositional Logic from the Viewpoint of Analytic Tableaux**

# Chapter I

# Preliminaries

## § 0. Foreword on Trees

Trees will play an important role throughout this work, so we shall commence with some pertinent definitions:

By an *unordered tree*, $\mathscr{T}$, we shall mean a collection of the following items:

(1) A set $S$ of elements called *points*.

(2) A function, $\ell$, which assigns to each point $x$ a positive integer $\ell(x)$ called the *level* of $x$.

(3) A relation $x R y$ defined in $S$, which we read "$x$ is a *predecessor* of $y$" or "$y$ is a *successor* of $x$". This relation must obey the following conditions:

$C_1$: There is a unique point $a_1$ of level 1. This point we call the *origin* of the tree.

$C_2$: Every point other than the origin has a unique predecessor.

$C_3$: For any points $x$, $y$, if $y$ is a successor of $x$, then $\ell(y) = \ell(x) + 1$.

We shall call a point $x$ an *end point* if it has no successors; a *simple point* if it has exactly one successor, and a *junction point* if it has more than one successor. By a *path* we mean any finite or denumerable sequence of points, beginning with the origin, which is such that each term of the sequence (except the last, if there is one) is the predecessor of the next. By a *maximal path* or *branch* we shall mean a path whose last term is an end point of the tree, or a path which is infinite.

It follows at once from $C_1$, $C_2$, $C_3$ that for any point $x$, there exists a unique path $P_x$ whose last term is $x$. If $y$ lies on $P_x$, then we shall say that $y$ *dominates* $x$, or that $x$ is *dominated* by $y$. If $x$ dominates $y$ and $x \neq y$, then we shall say that $x$ is (or lies) *above* $y$, or that $y$ lies *below* $x$. We shall say that $x$ is *comparable* with $y$ if $x$ dominates $y$ or $y$ dominates $x$. We shall say that $y$ is *between* $x$ and $z$ if $y$ is above one of the pair $\{x, z\}$ and below the other.

By an *ordered tree*, $\mathscr{T}$, we shall mean an unordered tree together with a function $\theta$ which assigns to each junction point $z$ a sequence $\theta(z)$ which contains no repetitions, and whose set of terms consists of all the successors of $z$. Thus, if $z$ is a junction point of an ordered tree, we can speak of the $1^{st}, 2^{nd}, \ldots, n^{th}, \ldots$ successors of $z$ (for any $n$ up to the number of successors of $z$) meaning, of course, the $1^{st}, 2^{nd}, \ldots, n^{th}, \ldots$ terms of $\theta(z)$.

For a simple point $x$, we shall also speak of the successor of $x$ as the *sole* successor of $x$.

We shall usually display ordered trees by placing the origin at the top and the successor(s) of each point $x$ below $x$, and in the order, from left to right, in which they are ordered in the tree. And we draw a line segment from $x$ to $y$ to signify that $y$ is a successor of $x$.

We shall have occasion to speak of adding "new" points as successors of an end point $x$ of a given tree $\mathcal{T}$. By this we mean more precisely the following: For any element $y$ outside $\mathcal{T}$, by the adjunction of $y$ as the sole successor of $x$, we mean the tree obtained by adding $y$ to the set $S$, and adding the ordered pair $\langle x, y \rangle$ to the relation $R$ (looked at as a set of ordered pairs), and extending the function $\ell$ by defining $\ell(y) = \ell(x) + 1$. For any distinct elements $y_1, \ldots, y_n$, each outside $S$, by the adjunction of $y_1, \ldots, y_n$ as respective $1^{st}, 2^{nd}, \ldots, n^{th}$ successors of $x$, we mean the tree obtained by adding the $y_i$ to $S$, adding the pairs $\langle x, y_i \rangle$ to $R$ and extending $\ell$ by setting $\ell(y_1) = \ldots = \ell(y_n) = \ell(x) + 1$, and extending the function $\theta$ by defining $\theta(x)$ to be the sequence $(y_1, \ldots, y_n)$. [It is obvious that the extended structure obtained is really a tree].

A tree is called *finitely generated* if each point has only finitely many successors. A tree, $\mathcal{T}$, is called *finite* if $\mathcal{T}$ has only finitely many points, otherwise the tree is called *infinite*. Obviously, a finitely generated tree may be infinite.

We shall be mainly concerned with ordered trees in which each junction point has exactly 2 successors. Such trees are called *dyadic* trees. For such trees we refer to the first successor of a junction point as the *left successor*, and the second successor as the *right successor*.

[*Exercise:* In a dyadic tree, define $x$ to be to the left of $y$ if there is a junction point whose left successor dominates $x$ and whose right successor dominates $y$. Prove that if $x$ is to the left of $y$ and $y$ is to the left of $z$, then $x$ is to the left of $z$].

## § 1. Formulas of Propositional Logic

We shall use for our undefined logical connectives the following 4 symbols:

(1) $\sim$ [read "not"],     (2) $\wedge$ [read "and"],
(3) $\vee$ [read "or"],     (4) $\supset$ [read "implies"].

These symbols are respectively called the *negation, conjunction, disjunction,* and *implication* symbols. The last 3 are collectively called *binary* connectives, the first ($\sim$) the *unary* connective.

Other symbols shall be:

(i) A denumerable set $p_1, p_2, \ldots, p_n, \ldots$ of symbols called *propositional variables*.

(ii) The two symbols (,), respectively called the *left parenthesis* and the *right parenthesis* (they are used for purposes of punctuation). Until we come to *First-Order Logic*, we shall use the word "variable" to mean *propositional variable*.

We shall use the letters "$p$", "$q$", "$r$", "$s$" to stand for any of the variables $p_1, p_2, \ldots, p_n, \ldots$. The notion of *formula* is given by the following recursive rules, which enable us to obtain new formulas from those already constructed:

$F_0$: Every propositional variable is a formula.

$F_1$: If $A$ is a formula so is $\sim A$.

$F_2, F_3, F_4$: If $A$, $B$ are formulas so are $(A \wedge B)$, $(A \vee B)$, $(A \supset B)$.

This recursive definition of "formula" can be made explicit as follows. By a *formation sequence* we shall mean any finite sequence such that each term of the sequence is either a propositional variable or is of the form $\sim A$, where $A$ is an earlier term of the sequence, or is of one of the forms $(A \wedge B)$, $(A \vee B)$, $(A \supset B)$, where $A$, $B$ are earlier terms of the sequence. Now we can define $A$ to be a *formula* if there exists a formation sequence whose last term is $A$. And such a sequence is also called a *formation sequence* for $A$.

For any formula $A$, by the *negation* of $A$ we mean $\sim A$. It will sometimes prove notationally convenient to write $A'$ in place of $\sim A$. For any 2 formulas $A$, $B$, we refer to $(A \wedge B)$, $(A \vee B)$, $(A \supset B)$ as the *conjunction, disjunction, conditional* of $A$, $B$ respectively. In a conditional formula $(A \supset B)$, we refer to $A$ as the *antecedent* and $B$ as the *consequent*.

We shall use the letters "$A$", "$B$", "$C$", "$X$", "$Y$", "$Z$" to denote formulas. We shall use the symbol "$b$" to denote any of the binary connectives $\wedge$, $\vee$, $\supset$; and when "$b$" respectively denotes $\wedge$, $\vee$, $\supset$ then $(X b Y)$ shall respectively mean $(X \wedge Y)$, $(X \vee Y)$, $(X \supset Y)$. We can thus state the formation rules more succinctly as follows:

$F_0$: Every propositional variable is a formula.

$F_1$: If $X$ is a formula so is $\sim X$.

$F_2$: If $X$, $Y$ are formulas, then for each of the binary connectives $b$, the expression $(X b Y)$ is a formula.

In displaying formulas by themselves (i.e. not as parts of other formulas), we shall omit outermost parentheses (since no ambiguity can result). Also, for visual perspicuity, we use square brackets [ ] interchangeably with parentheses, and likewise braces { }. Usually we shall use square brackets as exterior to parentheses, and braces as exterior to square brackets.

*Example.* Consider the following formula:

$$((((p \supset q) \wedge (q \vee r)) \supset (p \vee r)) \supset \sim (q \vee s)).$$

It is easier to read if displayed as follows:

$$\{[(p \supset q) \wedge (q \vee r)] \supset (p \vee r)\} \supset \sim (q \vee s).$$

*Biconditional* – we use "$X \leftrightarrow Y$" as an abbreviation for $(X \supset Y) \wedge (Y \supset X)$. The formula $X \leftrightarrow Y$ is called the *biconditional* of $X$, $Y$. It is read "$X$ if and only if $Y$" or "$X$ is equivalent to $Y$".

**Uniqueness of Decomposition.** It can be proved that every formula can be formed in only one way—i. e. for every formula $X$, one *and only one* of the following conditions holds:

(1)  $X$ is a propositional variable.

(2)  There is a *unique* formula $Y$ such that $X = Y'$.

(3)  There is a *unique* pair $X_1$, $X_2$ and a *unique* binary connective $b$ such that $X = (X_1 b X_2)$.

Thus no conjunction can also be a disjunction, or a conditional; no disjunction can also be a conditional. Also none of these can also be a negation. And, e. g., $(X_1 \wedge X_2)$ can be identical with $(Y_1 \wedge Y_2)$ only if $X_1 = Y_1$ and $X_2 = Y_2$ (and similarly with the other binary connectives). We shall not prove this here; perfectly good proofs can be found, e. g. in CHURCH [1] or KLEENE [1].

In our discussion below, we shall consider a more abstract approach in which this combinatorial lemma can be circumvented.

**\*Discussion.** First we wish to mention that some authors prefer the following formation rules for formulas:

$F_0'$:  Same as $F_0$.

$F_1'$:  If $X$ is a formula, so is $\sim (X)$.

$F_2'$:  If $X$, $Y$ are formulas, so is $(X) b (Y)$.

This second set of rules has the advantage of eliminating, at the outset, outermost parentheses, but has the disadvantage of needlessly putting parentheses around variables.

It seems to us that the following set of formation rules, though a bit more complicated to state, combines the advantages of the two preceding formulations, and involves using neither more nor less parentheses than is necessary to prevent ambiguity:

$F_0''$:  Same as before.

$F_1''$:  If $X$ is a formula but not a propositional variable and $p$ is a propositional variable, $\sim (X)$ and $\sim p$ are formulas.

$F_2''$: If $X$, $Y$ are both formulas, but neither $X$ nor $Y$ is a propositional variable, and if $p$, $q$ are propositional variables, then the following expressions are all formulas:

    (a) $(X)b(Y)$,
    (b) $(X)bq$,
    (c) $pb(Y)$,
    (d) $pbq$.

In all the above 3 approaches, one needs to prove the unique decomposition lemma for many subsequent results. Now let us consider yet another scheme (of a radically different sort) which avoids this.

First of all, we delete the parentheses from our basic symbols. We now define the *negation* of $X$, not as the symbol $\sim$ followed by the first symbol of $X$, followed by the second symbol of $X$, etc. but simply as the *ordered pair* whose first term is "$\sim$" and whose second term is $X$. And we define the *conjunction* of $X$, $Y$ as the *ordered triple* whose first term is $X$, whose second term is "$\wedge$" and whose third term is $Y$. [In contrast, the conjunction of $X$ and $Y$, as previously defined, is a sequence of $n+m+3$ terms, where $n$, $m$ are the respective number of terms of $X$, $Y$. The "3" additional terms are due to the left parenthesis, right parenthesis and "$\wedge$"]. Similarly we define the disjunction (conditional) of $X$, $Y$ as the ordered triple $\langle X, b, Y \rangle$ where $b$ is the binary connective in question.

Under this plan, a formula is either a (propositional) variable, an ordered pair (if it is a negation) or an ordered triple. Now, no ordered pair can also be an ordered triple, and neither one can be a single symbol. Furthermore, an ordered pair uniquely determines its first and second elements, and an ordered triple uniquely determines its first, second and third elements. Thus the fact that a formula can be formed in "only one way" is now immediate.

We remark that with this plan, we can (and will) still *use* parentheses to describe formulas, but the parentheses are *not* parts of the formula. For example, we write $X \wedge (Y \vee Z)$ to denote the ordered triple whose first term is $X$, whose second term is "$\wedge$", and whose third term is itself the ordered triple whose first, second and third terms are respectively, $Y$, $\vee$, $Z$. But (under this plan) the parentheses themselves do not belong to the object language[1]) but only to our metalanguage[1]).

The reader can choose for himself his preferred notion of "formula", since subsequent developments will not depend upon the choice.

---

[1]) The term *object language* is used to denote the language talked *about* (in this case the set of formal expressions of propositional logic), and the term *metalanguage* is used to denote the language in which we are talking about the object language (in the present case English augmented by various common mathematical symbols).

*Subformulas.* The notion of *immediate subformula* is given explicitly by the conditions:

$I_0$: Propositional variables have no immediate subformulas.

$I_1$: $\sim X$ has $X$ as an immediate subformula and no others.

$I_2 - I_4$: The formulas $X \wedge Y$, $X \vee Y$, $X \supset Y$ have $X$, $Y$ as immediate subformulas and no others.

We shall sometimes refer to $X$, $Y$ respectively as the *left immediate subformula, right immediate subformula* of $X \wedge Y$, $X \vee Y$, $X \supset Y$.

The notion of *subformula* is implicitly defined by the rules:

$S_1$: If $X$ is an immediate subformula of $Y$, or if $X$ is identical with $Y$, then $X$ is a subformula of $Y$.

$S_2$: If $X$ is a subformula of $Y$ and $Y$ is a subformula of $Z$, then $X$ is a subformula of $Z$.

The above implicit definition can be made explicit as follows: $Y$ is a subformula of $Z$ iff (i.e. if and only if) there exists a finite sequence starting with $Z$ and ending with $Y$ such that each term of the sequence except the first is an immediate subformula of the preceding term.

The only formulas having no immediate subformulas are propositional variables. These are sometimes called *atomic* formulas. Other formulas are called *compound* formulas. We say that a variable $p$ *occurs* in a formula $X$, or that $p$ is one of the variables of $X$, if $p$ is a subformula of $X$.

*Degrees; Induction Principles.* To facilitate proofs and definitions by induction, we define the *degree* of a formula as the number of occurrences of logical connectives. Thus:

$D_0$: A variable is of degree 0.

$D_1$: If $X$ is of degree $n$, then $\sim X$ is of degree $n+1$.

$D_2 - D_4$: If $X$, $Y$ are of degrees $n_1, n_2$, then $X \wedge Y$, $X \vee Y$, $X \supset Y$ are each of degree $n_1 + n_2 + 1$.

*Example.*

$p \wedge (q \vee \sim r)$ is of degree 3.

$p \wedge (q \vee r)$ is of degree 2.

We shall use the principle of *mathematical induction* (or of *finite descent*) in the following form. Let $S$ be a set of formulas ($S$ may be finite or infinite) and let $P$ be a certain property of formulas which we wish to show holds for every element of $S$. To do this it suffices to show the following two conditions:

(1) Every element of $S$ of degree 0 has the property $P$.

(2) If some element of $S$ of degree $> 0$ fails to have the property $P$, then some element of $S$ of lower degree also fails to have property $P$.

Of course, we can also use (2) in the equivalent form:

(2)′ For every element $X$ of $S$ of positive degree, if all elements of $S$ of degree less than that of $X$ have property $P$, then $X$ also has property $P$.

*Formation Trees.* It is sometimes useful to display all the subformulas of a given formula $X$ in the form of a *dyadic tree* which we call a *formation tree* for $X$... which completely shows the pedigree of $X$. We start the tree with the formula $X$ at the origin, and each node of the tree which is not a propositional variable "branches" into its immediate subformulas. More precisely, a formation tree for $X$ is an ordered dyadic tree $\mathcal{T}$ whose points are formulas (or rather occurrences of formulas, since the same formula may have several different occurrences on the tree) and whose origin is (an occurrence of) $X$, and such that the following 3 conditions hold:

(i) Each end point is (an occurrence of) a propositional variable.

(ii) Each simple point is of the form $\sim Y$ and has (an occurrence of) $Y$ as its sole successor.

(iii) Each junction point is of the form $X \, b \, Y$ and has (occurrences of) $X, Y$ as respective left and right successors.

As an example the following is a formation tree for the formula $[(p \wedge q) \supset (\sim p \vee \sim \sim q)] \vee (q \supset \sim p)$:

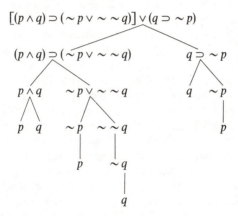

We might remark that the subformulas of a given formula $X$ are precisely those formulas which appear somewhere on the formation tree for $X$.

## § 2. Boolean Valuations and Truth Sets

Now we consider, in addition to the formulas of propositional logic, a set $\{t, f\}$ of two *distinct* elements, $t, f$. We refer to $t, f$ as *truth-values.* For any set $S$ of formulas, by a *valuation* of $S$, we mean a function $v$ from $S$ into the set $\{t, f\}$—i.e. a mapping which assigns to every element $X$ of $S$ one of the two values $t, f$. The value $v(X)$ of $X$ under $v$ is called the *truth value* of $X$ under $v$. We say that $X$ is *true under $v$* if $v(X) = t$, and *false under $v$* if $v(X) = f$.

Now we wish to consider valuations of the set $E$ of all formulas of propositional logic. We are not really interested in *all* valuations of $E$, but only in those which are "faithful" to the usual "truth-table" rules for the logical connectives. This idea we make precise in the following definition.

**Definition 1.** A valuation $v$ of $E$ is called a *Boolean* valuation if for every $X$, $Y$ in $E$, the following conditions hold:

$B_1$: The formula $\sim X$ receives the value $t$ if $X$ receives the value $f$ and $f$ if $X$ receives the value $t$.

$B_2$: The formula $X \wedge Y$ receives the value $t$ if $X$, $Y$ both receive the value $t$, otherwise $X \wedge Y$ receives the value $f$.

$B_3$: The formula $X \vee Y$ receives the value $t$ if at least one of $X$, $Y$ receives the value $t$, otherwise $X \vee Y$ receives the value $f$.

$B_4$: The formula $X \supset Y$ receives the value $f$ if $X$, $Y$ receive the respective values $t$, $f$, otherwise $X \supset Y$ receives the value $t$.

This concludes our definition of a Boolean valuation. We say that two valuations *agree* on a formula $X$ if $X$ is either true in both valuations or false in both valuations. And we say that 2 valuations agree on a *set $S$* of formulas if they agree on every element of the set $S$.

If $S_1$ is a subset of $S_2$ and if $v_1$, $v_2$ are respective valuations of $S_1$, $S_2$, then we say that $v_2$ is an *extension* of $v_1$ if $v_2$, $v_1$ agree on the smaller set $S_1$.

It is obvious that if 2 Boolean valuations agree on $X$ then they agree on $\sim X$ (why?), and if they agree on both $X$, $Y$ they must also agree on each of $X \wedge Y$, $X \vee Y$, $X \supset Y$ (why?). By mathematical induction it follows that if 2 Boolean valuations of $E$ agree on the set of all atomic elements of $E$ (i.e., on all propositional variables) then they agree on all of $E$. Stated otherwise, a valuation $v_0$ of the set of all *atomic* elements of $E$ can be extended to *at most* one Boolean valuation of $E$.

By an *interpretation* of a formula $X$ is meant an assignment of truth values to all of the variables which occur in $X$. More generally, by an interpretation of a set $W$ (of formulas) is meant an assignment of truth values to all the variables which occur in any of the elements of $W$. We can thus rephrase the last statement of the preceding paragraph by saying that any interpretation $v_0$ of $E$ can be extended to at most one Boolean valuation of $E$. That $v_0$ can be extended to at *least* one Boolean valuation of $E$ will be clear from the following considerations.

Consider a single formula $X$ and an interpretation $v_0$ of $X$—or for that matter any assignment $v_0$ of truth values to a set of propositional variables which include at least all variables of $X$ (and possibly others). It is easily verified by induction on the degree of $X$ that there exists one and only one way of assigning truth values to all *subformulas* of $X$ such that the *atomic* subformulas of $X$ (which are propositional variables) are assigned the same truth values as under $v_0$, and such that the truth value of each *compound* subformula $Y$ of $X$ is determined from the truth values of the immediate subformulas of $Y$ by the truth-table rules $B_1 - B_4$. [We might think of the situation as first constructing a formation tree for $X$, then assigning truth values to the end points in accordance with the interpretation $v_0$, and then working our way up the tree, successively assigning truth values to the junction and simple points, in terms of truth values already assigned to their successors, in accordance with the truth-table rules]. In particular, $X$ being a subformula of itself receives a truth value under this assignment; if this value is $t$ then we say that $X$ is true *under the interpretation* $v_0$, otherwise *false* under $v_0$. Thus we have now defined what it means for a formula $X$ to be true under an *interpretation*.

Now consider an interpretation, $v_0$, for the entire set $E$. Each element, $X$, of $E$ has a definite truth value under $v_0$ (in the manner we have just indicated); we let $v$ be that valuation which assigns to each element of $E$ its truth value under the interpretation $v_0$. The valuation $v$ is on the entire set $E$, and it is easily verified that $v$ is a Boolean valuation, and of course, $v$ is an extension of $v_0$. Thus it is indeed the case that every interpretation of $E$ can be extended to one (and only one) Boolean valuation of $E$.

*Tautologies.* The notion of *tautology* is perhaps the fundamental notion of propositional logic.

**Definition 2.** $X$ is a *tautology* iff $X$ is true in all Boolean valuations of $E$.

Equivalently, $X$ is a *tautology* iff $X$ is true under every *interpretation* of $E$. Now it is obvious that the truth value of $X$ under an interpretation of $E$ depends only on the truth values assigned to the variables which occur in $X$. Therefore, $X$ is a tautology if and only if $X$ is true under every interpretation of $X$. Letting $n$ be the number of variables which occur in $X$, there are exactly $2^n$ distinct interpretations of $X$. Thus the task of determining whether $X$ is or is not a tautology is purely a finite and mechanical one—just evaluate its truth value under each of its $2^n$ interpretations (which is tantamount to the familiar truth-table analysis).

**Definition 3.** A formula $X$ is called (truth-functionally) *satisfiable* iff $X$ is true in at least one Boolean valuation. A set $S$ of formulas is said to be (simultaneously) truth-functionally *satisfiable* iff there exists at least one Boolean valuation in which every element of $S$ is true. Such a valuation is said to *satisfy* $S$.

**Definition 4.** A set $S$ *truth-functionally implies* a formula $X$, or $X$ is *truth-functionally implied* by $S$, or is a *truth-functional consequence* of $S$ if $X$ is true in every Boolean valuation which satisfies $S$. We also say that $Y$ is truth-functionally implied by $X$ if $Y$ is truth functionally implied by the unit set $\{X\}$ ... i. e. if $Y$ is true in every Boolean valuation in which $X$ is true.

**Definition 5.** Two formulas $X$, $Y$ are called *truth functionally equivalent* iff $X$, $Y$ are true in the same Boolean valuations. [The reader should note that $X$ truth-functionally implies $Y$ iff $X \supset Y$ is a tautology, and that $X$ is truth-functionally equivalent to $Y$ iff the formula $X \leftrightarrow Y$ is a tautology].

*Truth Sets.* Let $v$ be a Boolean valuation, and let $S$ be the set of all formulas which are *true* under $v$. It is immediate from the definition of a Boolean valuation that the set $S$ obeys the following conditions (for every $X$, $Y$):

$S_1$: Exactly one of the pair $(X, \sim X)$ belongs to $S$. Stated otherwise $(\sim X) \in S$ iff $X \notin S$.

$S_2$: $(X \wedge Y)$ is in $S$ iff $X$, $Y$ are both in $S$.

$S_3$: $(X \vee Y)$ is in $S$ iff $X \in S$ or $Y \in S$.

$S_4$: $(X \supset Y)$ is in $S$ iff $X \notin S$ or $Y \in S$.

A set $S$ obeying the above conditions will be called *saturated* or will be said to be a *truth set*. Thus for any Boolean valuation, the set of all sentences true under the valuation is saturated. Indeed, if $v$ is an arbitrary valuation, and if $S$ is the set of all sentences which are true under $v$, then the following 2 conditions are *equivalent*:

(1) $v$ is a Boolean valuation,

(2) $S$ is *saturated*.

Now suppose that we start with a set $S$, and we define $v_s$ to be that valuation which assigns $t$ to every member of $S$, and $f$ to every formula outside $S$. [The function $v_s$ is sometimes referred to as the *characteristic function* of the set $S$.] It is again obvious that $S$ is *saturated* iff $v_s$ is a Boolean valuation.

Now the set of all sentences true under $v_s$ is obviously $S$ itself. Thus a set is saturated iff it is the set of all sentences true under some Boolean valuation. Thus a formula $X$ is a *tautology* iff it is an element of every truth set; stated otherwise, the set of tautologies is the *intersection* of all truth sets and a formula $X$ is *satisfiable* iff it is an element of some truth set. Stated otherwise, the set of satisfiable sentences is the *union* of all truth sets. Likewise a set $S$ truth-functionally implies $X$ iff $X$ belongs to every truth set which includes $S$.

We thus see that we really do not need to "import" these "foreign" elements $t$, $f$ in order to define our basic semantic notions. In some contexts it is technically more convenient to use $t$ and $f$ and Boolean valuations; in other it is simpler to use truth sets.

*Exercise* 1 [Truth Functional Equivalence]. We shall use "$\simeq$" in our *metalanguage* and write $X \simeq Y$ to mean that $X$ is *equivalent* to $Y$— i. e. that the formula $X \leftrightarrow Y$ is a tautology.

Now suppose that $X_1 \simeq X_2$. Prove the following equivalences:

$$\sim X_1 \simeq \sim X_2,$$

$$X_1 \wedge Y \simeq X_2 \wedge Y; \qquad Y \wedge X_1 \simeq Y \wedge X_2,$$
$$X_1 \vee Y \simeq X_2 \vee Y; \qquad Y \vee X_1 \simeq Y \vee X_2,$$
$$X_1 \supset Y \simeq X_2 \supset Y; \qquad Y \supset X_1 \simeq Y \supset X_2.$$

Using these facts, show that for any formula $Z$ which contains $X_1$ as a part, if we replace one or more occurrences of the part $X_1$ by $X_2$, the resulting formula is equivalent to $Z$.

*Exercise* 2 – [Important for Ch. XV!]. In some formulations of propositional logic, one uses "$t$", "$f$" as symbols of the object language itself; these symbols are then called *propositional constants*. And a *Boolean valuation* is redefined by adding the condition that $t$ must be given the value *truth* and $f$ *falsehood*. [Thus, e. g. $t$ by itself is a tautology; $f$ is unsatisfiable; $X \supset t$ is a tautology; $f \supset X$ is a tautology. Also, under any Boolean valuation $t \supset Y$ has the same truth value as $Y$; $X \supset f$ has the opposite value to $X$. Thus $t \supset Y$ is a tautology iff $Y$ is a tautology; $X \supset f$ is a tautology iff $X$ is unsatisfiable.]

Prove the following equivalences:

(1) $X \wedge t \simeq X$;   $X \wedge f \simeq f$,
(2) $X \vee t \simeq t$;   $X \vee f \simeq X$,
(3) $X \supset t \simeq t$;   $t \supset X \simeq X$,
(4) $X \supset f \simeq \sim X$;   $f \supset X \simeq t$,
(5) $\sim t \simeq f$;   $\sim f \simeq t$,
(6) $X \wedge Y \simeq Y \wedge X$;  $X \vee Y \simeq Y \vee X$.

Using these facts show that every formula $X$ with propositional constants is either equivalent to a formula $Y$ which contains no propositional constants or else it is equivalent to $t$ or to $f$.

*Exercise* 3. It is convenient to write a conjunction $(..(X_1 \wedge X_2) \wedge ... \wedge X_n)$ as $X_1 \wedge X_2 \wedge ... \wedge X_n$, and the formulas $X_1, X_2, ..., X_n$ are called the *components* of the conjunction. [Similarly we treat disjunctions.] By a *basic* conjunction is meant a conjunction with no repetitions of components such that each component is either a variable or the negation of a variable, but no variable and its negation are both components. [As an example, $p_1 \wedge \sim p_2 \wedge p_3$ is a basic conjunction—so is $\sim p_1 \wedge p_2 \wedge \sim p_3$— so is $\sim p_1 \wedge p_2 \wedge p_3$.] By a *disjunctive normal formula* is mean a formula $C_1 \vee ... \vee C_k$, where each $C_i$ is a *basic* conjunction. [As an example the formula $(p_1 \wedge \sim p_2 \wedge p_3) \vee (\sim p_1 \wedge p_2 \wedge p_3) \vee (\sim p_1 \wedge p_2 \wedge p_3)$ is a disjunctive normal formula.] A disjunctive normal formula is also sometimes referred to as a formula in disjunctive normal form. If we allow propositional constants $t, f$ into our formal language, then the formula $f$ is also said to be a disjunctive normal formula.

Prove that every formula can be put into disjunctive normal form—i.e. is equivalent to some disjunctive normal formula. [Hint: Make a truth-table for the formula. Each line of the table which comes out "$T$" will yield one of the basic conjunctions of the disjunctive normal form.]

*Exercise 4.* A binary connective $C$ is said to be *definable* from connectives $C_1, \ldots, C_k$ if there exists a formula in two variables $p, q$ which uses just the connectives $C_1, \ldots, C_k$ and which is equivalent to the formula $p\,C\,q$.

As an example, $\vee$ is definable from $\{\sim, \wedge\}$, because the formula $\sim(\sim p \wedge \sim q)$ is equivalent to $p \vee q$.

> Prove:   $\wedge$ is definable from $\{\sim, \vee\}$,
>          $\supset$ is definable from $\{\sim, \wedge\}$,
>          $\supset$ is definable from $\{\sim, \vee\}$,
>          $\wedge$ is definable from $\{\sim, \supset\}$,
>          $\vee$ is definable from $\{\sim, \supset\}$.

*Exercise 5.* Let us introduce Sheffer's *stroke* symbol "$|$" as a binary connective for propositional logic, and add the formation rule "If $X$, $Y$ are formulas, so is $(X|Y)$". [We read "$X|Y$" as "$X$ is *incompatible* with $Y$" or "either $X$ or $Y$ is false".] A Boolean valuation is then re-defined by adding the conditions "$X|Y$ is true under $v$ iff at least one of $X$, $Y$ is false under $v$":

(a) Show that $|$ is definable from the other connectives.

(b) Show that *all* the other connectives are definable from $|$ ($\sim$ is definable from $|$ in the sense that there is a formula $\varphi(p)$ involving just the stroke connective and one propositional variable $p$ such that $\varphi(p)$ is equivalent to $\sim p$).

Do the same for the *joint denial connective* $\downarrow$ (where $X \downarrow Y$ is read "both $X$, $Y$ are false". Show that all other connectives are definable from $\downarrow$.

It can be shown that $|$, $\downarrow$ are the *only* binary connectives which each suffice to define all other connectives. [This is not easy! The "virtuoso" reader might wish to try his hand at this as an exercise.]

## Chapter II

# Analytic Tableaux

We now describe an extremely elegant and efficient proof procedure for propositional logic which we will subsequently extend to first order logic, and which shall be basic to our entire study. This method, which we term *analytic tableaux*, is a variant of the "semantic tableaux" of Beth [1], or of methods of Hintikka [1]. (Cf. also Anderson and Belnap [1].) Our present formulation is virtually that which we introduced in [1]. Ultimately, the whole idea derives from Gentzen [1], and we shall subsequently study the relation of analytic tableaux to the original methods of Gentzen.

## § 1. The Method of Tableaux

We begin by noting that under any interpretation the following eight facts hold (for any formulas $X, Y$):

1) a) If $\sim X$ is true, then $X$ is false.
   b) If $\sim X$ is false, then $X$ is true.
2) a) If a conjunction $X \wedge Y$ is true, then $X, Y$ are both true.
   b) If a conjunction $X \wedge Y$ is false, then either $X$ is false or $Y$ is false.
3) a) If a disjunction $X \vee Y$ is true, then either $X$ is true or $Y$ is true.
   b) If a disjunction $X \vee Y$ is false, then both $X, Y$ are false.
4) a) If $X \supset Y$ is true, then either $X$ is false or $Y$ is true.
   b) If $X \supset Y$ is false, then $X$ is true and $Y$ is false.

These eight facts provide the basis of the tableau method.

*Signed Formulas.* At this stage it will prove useful to introduce the symbols "$T$", "$F$" to our object language, and define a *signed* formula as an expression $TX$ or $FX$, where $X$ is a (unsigned) formula. (Informally, we read "$TX$" as "$X$ is true" and "$FX$" as "$X$ is false".)

**Definition.** Under any interpretation, a signed formula $TX$ is called *true* if $X$ is true, and *false* if $X$ is false. And a signed formula $FX$ is called *true* if $X$ is false, and *false* if $X$ is true.

Thus the truth value of $TX$ is the same as that of $X$; the truth value of $FX$ is the same as that of $\sim X$.

By the *conjugate* of a signed formula we mean the result of changing "$T$" to "$F$" or "$F$" to "$T$" (thus the conjugate of $TX$ is $FX$; the conjugate of $FX$ is $TX$).

*Illustration of the Method of Tableaux.* Before we state the eight rules for the construction of tableaux, we shall illustrate the construction with an example.

Suppose we wish to prove the formula $[p \vee (q \wedge r)] \supset [(p \vee q) \wedge (p \vee r)]$. The following is a tableau which does this; the explanation is given immediately following the tableau:

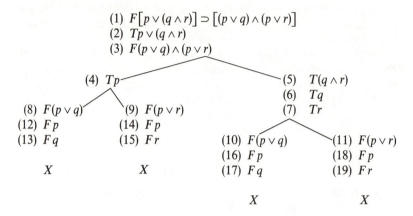

*Explanation.* The tableau was constructed as follows. We see if we can derive a contradiction from the assumption that the formula $[p \vee (q \wedge r)] \supset [(p \vee q) \wedge (p \vee r)]$ is false. So our first line consists of this formula preceded by the letter "*F*". Now, a formula of the form $X \supset Y$ can be false only if $X$ is true and $Y$ is false. (Cf. condition $B_4$ of a Boolean valuation.) Thus (in the language of tableaux) $TX$ and $FY$ are *direct* consequences of the (signed) formula $F(X \supset Y)$. So we write the lines (2) and (3) as *direct* consequences of line (1). Now let us look at line (2); it is of the form $T(X \vee Y)$ (where $X = p$, $Y = (q \wedge r)$.) We can not draw any *direct* conclusion about the truth value of $X$ nor about the truth value of $Y$; all we can infer is that *either* $TX$ *or* $TY$. So the tableau *branches* into two columns; one for each possibility. Thus line (2) *branches* into lines (4) and (5). Line (5), viz. $T(q \wedge r)$ immediately yields $Tq$ and $Tr$ as direct consequences; we thus have lines (6) and (7). Now look at (3). It is of the form $F(X \wedge Y)$. This means that *either* $FX$ *or* $FY$. We also know that either (4) or (5) holds. So for *each* of the possibilities (4), (5) we have one of the two possibilities $FX$, $FY$. There are hence now four possibilities. So each of the branches (4), (5) branches again into the possibilities $FX$, $FY$. More specifically, (4) branches to (8), (9), and (5) branches to (10), (11) (which are respectively the same as (8), (9)). Lines (12), (13) are direct consequences of (8); (14), (15) are direct consequences of (9); (16), (17) of (10); and (18), (19) of (11).

We now look at the leftmost branch and we shall see that (12) is a direct contradiction of (4) (i. e. it is the conjugate of (4)), so we put a cross after (13) to signify that this branch leads to a contradiction. Similarly, (14) contradicts (4), so we "close" the branch leading to (15)—i.e. we put a cross after (15). The next branch is closed by virtue of (17) and (6). Finally, the rightmost branch is closed by virtue of (19) and (7). Thus all branches lead to a contradiction, so line (1) is untenable. Thus $[p \vee (q \wedge r)] \supset [(p \vee q) \wedge (p \vee r)]$ can never be false in any interpretation, so it is a tautology.

**Remarks.** (i) The numbers put to the left of the lines were only for the purpose of identification in the above explanations; we do not need them for the actual construction.

(ii) We could have closed some of our branches a bit earlier; lines (13), (15) are superfluous. In subsequent examples we shall close a branch as soon as a contradiction appears (a contradiction that is of the form of two formulas $FX$, $TX$).

*Rules for the Construction of Tableaux.* We now state all the rules in schematic form; explanations immediately follow. For each logical connective there are two rules; one for a formula preceded by "$T$", the other for a formula preceded by "$F$":

$$
1)\quad \frac{T \sim X}{F X} \qquad \frac{F \sim X}{T X}
$$

$$
2)\quad \frac{T(X \wedge Y)}{\begin{array}{c} T X \\ T Y \end{array}} \qquad \frac{F(X \wedge Y)}{F X \,|\, F Y}
$$

$$
3)\quad \frac{T(X \vee Y)}{T X \,|\, T Y} \qquad \frac{F(X \vee Y)}{\begin{array}{c} F X \\ F Y \end{array}}
$$

$$
4)\quad \frac{T(X \supset Y)}{F X \,|\, T Y} \qquad \frac{F(X \supset Y)}{\begin{array}{c} T X \\ F Y \end{array}}
$$

*Some Explanations.* Rule 1) means that from $T \sim X$ we can directly infer $FX$ (in the sense that we can subjoin $FX$ to any branch passing through $T \sim X$) and that from $F \sim X$ we can directly infer $TX$. Rule 2) means that $T(X \wedge Y)$ directly yields both $TX$, $TY$, whereas $F(X \wedge Y)$ *branches* into $FX, FY$. Rules 3) and 4) can now be understood analogously.

Signed formulas, other than signed variables, are of two types; (A) those which have *direct* consequences (viz. $F \sim X$, $T \sim X$, $T(X \wedge Y)$, $F(X \vee Y)$, $F(X \supset Y)$); (B) those which *branch* (viz. $F(X \wedge Y)$, $T(X \vee Y)$, $T(X \supset Y)$).

It is practically desirable in constructing a tableau, that when a line of type (A) appears on the tableau, we simultaneously subjoin its consequences to *all* branches which pass through that line. Then that line need never be used again. And in using a line of type (B), we divide *all* branches which pass through that line into sub-branches, and the line need never be used again. For example, in the above tableau, we use (1) to get (2) and (3), and (1) is never used again. From (2) we get (4) and (5), and (2) is never used again. Line (3) yields (8), (9), (10), (11) and (3) is never used again, etc.

If we construct a tableau in the above manner, it is not difficult to see, that after a finite number of steps we must reach a point where every line has been used (except of course, for signed variables, which are never used at all to create new lines). At this point our tableau is *complete* (in a precise sense which we will subsequently define).

One way to complete a tableau is to work systematically downwards i.e. never to use a line until all lines above it (on the same branch) have been used. Instead of this procedure, however, it turns out to be more efficient to give priority to lines of type (A)—i.e. to use up all such lines at hand before using those of type (B). In this way, one will omit repeating the same formula on different branches; rather it will have only one occurrence *above* all those branch points.

As an example of both procedures, let us prove the formula $[p \supset (q \supset r)] \supset [(p \supset q) \supset (p \supset r)]$. The first tableau works systematically downward; the second uses the second suggestion. For the convenience of the reader, we put to the right of each line the number of the line from which it was inferred.

### First Tableau

$(1)\ F[p \supset (q \supset r)] \supset [(p \supset q) \supset (p \supset r)]$

$(2)\ Tp \supset (q \supset r)\ (1)$

$(3)\ F(p \supset q) \supset (p \supset r)\ (1)$

| | | | | |
|---|---|---|---|---|
| $(4)\ Fp\ (2)$ | | $(5)\ T(q \supset r)\ (2)$ | | |
| $(6)\ T(p \supset q)\ (3)$ | | $(8)\ T(p \supset q)\ (3)$ | | |
| $(7)\ F(p \supset r)\ (3)$ | | $(9)\ F(p \supset r)\ (3)$ | | |

$(10)\ Fp\ (6)$ | $(11)\ Tq\ (6)$     $(14)\ Fq\ (5)$        $(15)\ Tr\ (5)$

$(12)\ Tp\ (7)$ | $(13)\ Tp\ (7)$   $(16)\ Fp\ (8)$ | $(17)\ Tq\ (8)$   $(18)\ Fp\ (8)$ | $(19)\ Tq\ (8)$

$X$       $X$      $(20)\ Tp\ (9)$ | $X$     $(21)\ Tp\ (9)$ | $(22)\ Tp\ (9)$

                                $X$                  $X$    | $(23)\ Fr\ (9)$

                                                                  $X$

### Second Tableau

(1) $F[p \supset (q \supset r)] \supset [(p \supset q) \supset (p \supset r)]$

(2) $Tp \supset (q \supset r)$ (1)

(3) $F(p \supset q) \supset (p \supset r)$ (1)

(4) $T(p \supset q)$ (3)

(5) $F(p \supset r)$ (3)

(6) $Tp$ (5)

(7) $Fr$ (5)

$$
\begin{array}{c|ccc}
(8)\ Fp\ (2) & (9)\ T(q \supset r)\ (2) & & \\
X & (10)\ Fp\ (4) & (11)\ Tq\ (4) & \\
& X & (12)\ Fq\ (9)\ \mid\ (13)\ Tr\ (9) \\
& & X \qquad\qquad X
\end{array}
$$

It is apparent that Tableau (2) is quicker to construct than Tableau (1), involving only 13 rather than 23 lines.

As another practical suggestion, one might put a check mark to the right of a line as soon as it has been used. This will subsequently aid the eye in hunting upward for lines which have not yet been used. (The check marks may be later erased, if the reader so desires.)

The method of analytic tableaux can also be used to show that a given formula is a truth functional consequence of a given finite set of formulas. Suppose we wish to show that $X \supset Z$ is a truth-functional consequence of the two formulas $X \supset Y$, $Y \supset Z$. We could, of course, simply show that $[(X \supset Y) \wedge (Y \supset Z)] \supset (X \supset Z)$ is a tautology. Alternatively, we can construct a tableau starting with

$$T(X \supset Y),$$
$$T(Y \supset Z),$$
$$F(X \supset Z)$$

and show that all branches close.

In general, to show that $Y$ is truth-functionally implied by $X_1,...,X_n$, we can construct either a closed analytic tableau starting with $F(X_1 \wedge ... \wedge X_n) \supset Y$, or one starting with

$$TX_1$$
$$\vdots$$
$$TX_n$$
$$FY$$

*Tableaux using unsigned formulas.* Our use of the letters "$T$" and "$F$", though perhaps heuristically useful, is theoretically quite dispensable—simply delete every "$T$" and substitute "$\sim$" for "$F$". (In which case, incidentally, the first half of Rule 1) becomes superfluous.) The rules then become:

$$1) \quad \frac{\sim \sim X}{X}$$

$$2) \quad \frac{X \wedge Y}{\begin{array}{c} X \\ Y \end{array}} \qquad \frac{\sim (X \wedge Y)}{\sim X \mid \sim Y}$$

$$3) \quad \frac{X \vee Y}{X \mid Y} \qquad \frac{\sim (X \vee Y)}{\begin{array}{c} \sim X \\ \sim Y \end{array}}$$

$$4) \quad \frac{X \supset Y}{\sim X \mid Y} \qquad \frac{\sim (X \supset Y)}{\begin{array}{c} X \\ \sim Y \end{array}}$$

In working with tableaux which use unsigned formulas, "closing" a branch naturally means terminating the branch with a cross, as soon as two formulas appear, one of which is the *negation* of the other. A tableau is called *closed* if every branch is closed.

By a tableau *for* a formula $X$, we mean a tableau which starts with $X$. If we wish to prove a formula $X$ to be a tautology, we construct a tableau not for the formula $X$, but for its negation $\sim X$.

*A Unifying Notation.* It will save us considerable repetition of essentially the same arguments in our subsequent development if we use the following unified notation which we introduced in [2].

We use the letter "$\alpha$" to stand for any signed formula of type A—i.e. of one of the five forms $T(X \wedge Y)$, $F(X \vee Y)$, $F(X \supset Y)$, $T \sim X$, $F \sim X$. For every such formula $\alpha$, we *define* the two formulas $\alpha_1$ and $\alpha_2$ as follows:

$$\begin{array}{ll}
\text{If } \alpha = T(X \wedge Y), & \text{then } \alpha_1 = TX \text{ and } \alpha_2 = TY. \\
\text{If } \alpha = F(X \vee Y), & \text{then } \alpha_1 = FX \text{ and } \alpha_2 = FY. \\
\text{If } \alpha = F(X \supset Y), & \text{then } \alpha_1 = TX \text{ and } \alpha_2 = FY. \\
\text{If } \alpha = T \sim X, & \text{then } \alpha_1 = FX \text{ and } \alpha_2 = FX. \\
\text{If } \alpha = F \sim X, & \text{then } \alpha_1 = TX \text{ and } \alpha_2 = TX.
\end{array}$$

For perspicuity, we summarize these definitions in the following table:

| $\alpha$ | $\alpha_1$ | $\alpha_2$ |
|---|---|---|
| $T(X \wedge Y)$ | $TX$ | $TY$ |
| $F(X \vee Y)$ | $FX$ | $FY$ |
| $F(X \supset Y)$ | $TX$ | $FY$ |
| $T \sim X$ | $FX$ | $FX$ |
| $F \sim X$ | $TX$ | $TX$ |

We note that under any interpretation, $\alpha$ is true iff $\alpha_1$, $\alpha_2$ are *both* true. Accordingly, we shall also refer to an $\alpha$ as a formula of *conjunctive* type. We use "$\beta$" to stand for any signed formula of type $B$—i.e. one of the three forms $F(X \wedge Y)$, $T(X \vee Y)$, $T(X \supset Y)$. For every such formula $\beta$, we define the two formulas $\beta_1$, $\beta_2$ as in the following table:

| $\beta$ | $\beta_1$ | $\beta_2$ |
|---|---|---|
| $F(X \wedge Y)$ | $FX$ | $FY$ |
| $T(X \vee Y)$ | $TX$ | $TY$ |
| $T(X \supset Y)$ | $FX$ | $TY$ |

In any interpretation, $\beta$ is true iff *at least one* of the pair $\beta_1$, $\beta_2$ is true. Accordingly, we shall refer to any $\beta$-type formula as a formula of *disjunctive* type.

We shall sometimes refer to $\alpha_1$ as the *first component* of $\alpha$ and $\alpha_2$ as the *second component* of $\alpha$. Similarly, for $\beta$.

By the *degree* of a signed formula $TX$ or $FX$ we mean the degree of $X$. We note that $\alpha_1$, $\alpha_2$ are each of lower degree than $\alpha$, and $\beta_1$, $\beta_2$ are each of lower degree than $\beta$. Signed variables, of course, are of degree 0.

We might also employ an $\alpha$, $\beta$ classification of *unsigned* formulas in an analogous manner, simply delete all "$T$", and replace "$F$" by "$\sim$". The tables would be as follows:

| $\alpha$ | $\alpha_1$ | $\alpha_2$ |
|---|---|---|
| $X \wedge Y$ | $X$ | $Y$ |
| $\sim(X \vee Y)$ | $\sim X$ | $\sim Y$ |
| $\sim(X \supset Y)$ | $X$ | $\sim Y$ |
| $\sim \sim X$ | $X$ | $X$ |

| $\beta$ | $\beta_1$ | $\beta_2$ |
|---|---|---|
| $\sim(X \wedge Y)$ | $\sim X$ | $\sim Y$ |
| $X \vee Y$ | $X$ | $Y$ |
| $X \supset Y$ | $\sim X$ | $Y$ |

Let us now note that whether we work with signed or unsigned formulas, all our tableau rules can be succinctly lumped into the following two:

$$\text{Rule } A - \quad \frac{\alpha}{\begin{array}{c} \alpha_1 \\ \alpha_2 \end{array}} \qquad\qquad \text{Rule } B - \quad \frac{\beta}{\beta_1 \mid \beta_2}$$

N.B. In working with signed formulas, there arise situations in which it is better to regard signed formulas of the form $T \sim X$ or $F \sim X$ as of both the $\alpha$-type *and* the $\beta$-type. If we agree to let $\beta$ be such a formula, it still remains true that under any interpretation, $\beta$ is true iff at least one of $\beta_1, \beta_2$ is true (for in such a case, $\beta_1$ and $\beta_2$ are identical expressions). Under this extended use of "$\beta$", Rule $B$ when applied to an expression $T \sim X$ or $F \sim X$ would yield two identical branches. So in practice, we need consider only one. Indeed, in practice we need consider only one branch even in other cases when $\beta_1$ and $\beta_2$ are identical— e.g. if $\beta = T(X \vee X)$ it would be pointless to divide the tableau into two branches, $TX$ and $TX$!

*Some Properties of Conjugation.* One reason for the desirability of using "$\beta$" in the extended manner which we just discussed, is that our operation of conjugation obeys the following pleasant symmetric laws:

$J_0$: (a) $\bar{X}$ is distinct from $X$;
    (b) $\bar{\bar{X}} = X$.
$J_1$: (a) The conjugate of any $\alpha$ is some $\beta$.
    (b) The conjugate of any $\beta$ is some $\alpha$.
$J_2$: (a) $(\bar{\alpha})_1 = \overline{\alpha_1}$; $(\bar{\alpha})_2 = \overline{\alpha_2}$;
    (b) $(\bar{\beta})_1 = \overline{\beta_1}$; $(\bar{\beta})_2 = \overline{\beta_2}$.

Law $J_1 -$ (a) would fail under the original definition of $\beta$.

Law $J_2$ says that if we take any signed formula $X$ other than a signed variable, if we first conjugate $X$ and then take the first (second) component of this conjugate we obtain the same result as if we first take the first (respectively second) component of $X$ and then conjugate it. (For example, suppose $\alpha = F(X \supset Y)$. Then $\bar{\alpha} = T(X \supset Y)$, $(\bar{\alpha})_1 = FX$, $(\bar{\alpha})_2 = TY$. On the other hand, $\alpha_1 = TX$, $\alpha_2 = FY$, so $\bar{\alpha}_1 = FX$, $\bar{\alpha}_2 = TY$). The reader can verify the remaining four cases for $\alpha$ and similarly the cases for $\beta$.

We note that $J_2$ can be equivalently stated as follows: If $\alpha$ is the conjugate of $\beta$, then $\alpha_1$ is the conjugate of $\beta_1$ and $\alpha_2$ is the conjugate of $\beta_2$.

*Truth Sets re-visited.* For the moment let us work with unsigned formulas (interpreting "$\alpha$" and "$\beta$" accordingly). Let $S$ be a set of unsigned formulas. We leave it to the reader to verify that $S$ is a *truth* set (as defined in Chapter I) if and only if $S$ has the following three properties (for every $X, \alpha, \beta$):

(0) Exactly one of $X, \sim X$ belongs to $S$.

(A) $\alpha$ belongs to $S$ if and only if $\alpha_1, \alpha_2$ both belong to $S$.

(B) $\beta$ belongs to $S$ if and only if at least one of $\beta_1, \beta_2$ belong to $S$.

We shall also refer to a set $S$ of *signed* formulas as a *valuation* set or *truth set* if it obeys conditions $(A), (B)$ above and in place of (0), the condition "exactly one of $TX, FX$ belongs to $S$". We shall also refer to valuation sets of signed formulas as *saturated* sets.

*Exercise.* Show that if a set $S$ of unsigned formulas satisfies $(A)$, $(B)$ above and in place of (0) the weaker condition "for every propositional variable $p$, exactly one of $p$, $\sim p$ lies in $S$" then it follows that for *every* formula $X$, exactly one of $X$, $\sim X$ lies in $S$—i.e. that $S$ is then a truth set. State and prove an analogous result for signed formulas.

We have defined a set $S$ of signed formulas to be a truth set if it satisfies the laws:

(0) For any $X$, exactly one of $X$, $\bar{X}$ belongs to $S$.

(a) $\alpha \in S$ iff $\alpha_1 \in S$ and $\alpha_2 \in S$.

(b) $\beta \in S$ iff $\beta_1 \in S$ or $\beta_2 \in S$.

We wish to point out that $(b)$ is superfluous in the presence of (0) and $(a)$ and also that $(a)$ is superfluous in the presence of (0) and $(b)$. Let us prove that $(b)$ follows from (0) and $(a)$—in doing this it will shorten our work to use our laws $J_0 - J_2$ on conjugation. Assume now that $S$ satisfies conditions (0), $(a)$. We must show that $\beta$ belongs to $S$ iff $\beta_1$ belongs to $S$ or $\beta_2$ belongs to $S$. Suppose that $\beta \in S$. If neither $\beta_1$ nor $\beta_2$ belonged to $S$, then $\bar{\beta}_1$ and $\bar{\beta}_2$ would belong to $S$ (since by (0) at least one of $\beta_1$, $\bar{\beta}_1$ belongs to $S$ and at least one of $\beta_2$, $\bar{\beta}_2$ belongs to $S$). Now we use the fact that $\bar{\beta}$ is some $\alpha$, and $\bar{\beta}_1 = \alpha_1$, $\bar{\beta}_2 = \alpha_2$. So $\alpha_1, \alpha_2$ both belong to $S$. Then by $(a)$, $\alpha$ belongs to $S$—i.e. $\bar{\beta}$ belongs to $S$. This would mean that $\beta$, $\bar{\beta}$ both belong to $S$, contrary to condition (0). This proves the first half. Conversely, suppose at least one of $\beta_1, \beta_2$ belongs to $S$—let us say $\beta_1$ (a perfectly analogous argument works for $\beta_2$). If $\beta$ fails to belong to $S$, then $\bar{\beta}$ belongs to $S$. But $\bar{\beta}$ is some $\alpha$, hence by $(a)$, $(\bar{\beta})_1$ and $(\bar{\beta})_2$ belong to $S$—hence (by $J_2$) $\bar{\beta}_1$ belongs to $S$. This implies $\beta_1$ and $\bar{\beta}_1$ both belong to $S$, contrary to condition (0). This concludes the proof.

We leave it to the reader to verify that it is likewise possible to obtain $(a)$ from (0) and $(b)$. We also remark that the same result holds for truth sets of *unsigned* formulas (but the verification can not be carried out as elegantly, since we have no conjugation operation satisfying $J_0, J_1, J_2$; the proof must be ground out by considering all cases separately).

*Exercise.* Call a set *downward closed* if for every $\alpha$ and $\beta$, (1) if $\alpha$ is in $S$, then $\alpha_1, \alpha_2$ are both in $S$; (2) if $\beta$ is in $S$, then at least one of $\beta_1, \beta_2$ is in $S$. Call a set *upward closed* if the converse conditions hold—i.e. (1) if $\alpha_1, \alpha_2$ are both in $S$, so is $\alpha$; (2) if either $\beta_1$ or $\beta_2$ is in $S$, so is $\beta$. Show that any *downward* closed set satisfying condition (0) (viz. that for every $X$, exactly one of $X, \bar{X}$ is in $S$) is a truth set. Show that any *upward* closed set satisfying (0) is a truth set.

*Precise Definition of Tableaux.* We have deliberately waited until the introduction of our unified notation in order to give a precise definition of an analytic tableau, since the definition can now be given more compactly.

**Definition.** An analytic tableau for $X$ is an ordered dyadic tree, whose points are (occurrences of) formulas, which is constructed as follows. We start by placing $X$ at the origin. Now suppose $\mathcal{T}$ is a tableau for $X$ which has already been constructed; let $Y$ be an end point. Then we may extend $\mathcal{T}$ by either of the following two operations.

($A$) If some $\alpha$ occurs on the path $P_Y$, then we may adjoin either $\alpha_1$ or $\alpha_2$ as the sole successor of $Y$. (In practice, we usually successively adjoin $\alpha_1$ and then $\alpha_2$.)

($B$) If some $\beta$ occurs on the path $P_Y$, then we may simultaneously adjoin $\beta_1$ as the left successor of $Y$ and $\beta_2$ as the right successor of $Y$.

The above inductive definition of tableau for $X$ can be made explicit as follows. Given two ordered dyadic trees $\mathcal{T}_1$ and $\mathcal{T}_2$, whose points are occurrences of formulas, we call $\mathcal{T}_2$ a *direct extension* of $\mathcal{T}_1$ if $\mathcal{T}_2$ can be obtained from $\mathcal{T}_1$ by one application of the operation ($A$) or ($B$) above. Then $\mathcal{T}$ is a tableau for $X$ iff there exists a finite sequence $(\mathcal{T}_1, \mathcal{T}_2, \ldots, \mathcal{T}_n = \mathcal{T})$ such that $\mathcal{T}_1$ is a 1-point tree whose origin is $X$ and such that for each $i < n, \mathcal{T}_{i+1}$ is a direct extension of $\mathcal{T}_i$.

To repeat some earlier definitions (more or less informally stated) a branch $\theta$ of a tableau for signed (unsigned) formulas is *closed* if it contains some signed formula and its conjugate (or some unsigned formula and its negation, if we are working with unsigned formulas.) And $\mathcal{T}$ is called closed if every branch of $\mathcal{T}$ is closed. By a *proof* of (an unsigned formula) $X$ is meant a closed tableau for $FX$ (or for $\sim X$, if we work with unsigned formulas.)

*Exercise.* By the tableau method, prove the following tautologies:

(1) $q \supset (p \supset q)$

(2) $((p \supset q) \wedge (q \supset r)) \supset (p \supset r)$

(3) $((p \supset q) \wedge (p \supset r)) \supset (p \supset (q \wedge r))$

(4) $[((p \supset r) \wedge (q \supset r)) \wedge (p \vee q)] \supset r$

(5) $\sim (p \wedge q) \supset (\sim p \vee \sim q)$

(6) $\sim (p \vee q) \supset (\sim p \wedge \sim q)$

(7) $(\sim p \vee \sim q) \supset \sim (p \wedge q)$

(8) $(p \vee (q \wedge r)) \supset ((p \vee q) \wedge (p \vee r))$

## § 2. Consistency and Completeness of the System

*Consistency.* It is intuitively rather obvious that any formula provable by the tableau method must be a tautology—equivalently, given any closed tableau, the origin must be unsatisfiable. This intuitive conviction can be justified by the following argument.

Consider a tableau $\mathcal{T}$ and an interpretation $v_0$ whose domain includes at least all the variables which occur in any point of $\mathcal{T}$. Let us call a branch $\theta$ of $\mathcal{T}$ *true* under $v_0$ if *every* term of $\theta$ is true under $v_0$. And we shall say that the tableau $\mathcal{T}$ (as a whole) is true under $v_0$ iff at least one branch of $\mathcal{T}$ is true under $v_0$.

The next step is to note that if a tableau $\mathcal{T}_2$ is an immediate extension of $\mathcal{T}_1$, then $\mathcal{T}_2$ must be true in every interpretation in which $\mathcal{T}_1$ is true. For if $\mathcal{T}_1$ is true, it must contain at least one true branch $\theta$. Now $\mathcal{T}_2$ was obtained from $\mathcal{T}_1$ by adding one or two successors to the end point of some branch $\theta_1$ of $\mathcal{T}_1$; if $\theta_1$ is distinct from $\theta$, then $\theta$ is still a branch of $\mathcal{T}_2$, hence $\mathcal{T}_2$ contains the true branch $\theta$, so $\mathcal{T}_2$ is true. On the other hand, suppose $\theta$ is identical with $\theta_1$—i.e. suppose $\theta$ is the branch of $\mathcal{T}_1$ which was extended in $\mathcal{T}_2$. If $\theta$ was extended by operation $(A)$, then some $\alpha$ appears as a term in $\theta$, and $\theta$ has been extended either to $(\theta_1, \alpha_1)$ or to $(\theta_1, \alpha_2)$, so either $(\theta_1, \alpha_1)$ or $(\theta_1, \alpha_2)$ is a branch of $\mathcal{T}_2$. But $\alpha_1, \alpha_2$ are both true since $\alpha$ is, hence $\mathcal{T}_2$ contains the true branch $(\theta_1, \alpha_1)$ or $(\theta_1, \alpha_2)$. If $\theta$ was extended by operation $(B)$, then some $\beta$ occurs in $\theta$ and both $(\theta_1, \beta_1)$ and $(\theta_1, \beta_2)$ are branches of $\mathcal{T}_2$. But since $\beta$ is true, then at least one of $\beta_1, \beta_2$ is true, hence one of the branches $(\theta_1, \beta_1)$ or $(\theta_1, \beta_2)$ of $\mathcal{T}_2$ is true, so again $\mathcal{T}_2$ is true.

We have thus shown that any immediate extension of a tableau which is true (under a given interpretation) is again true (under the given interpretation). From this it follows by mathematical induction that for any tableau $\mathcal{T}$, if the origin is true under a given interpretation $v_0$, then $\mathcal{T}$ must be true under $v_0$. Now a closed tableau $\mathcal{T}$ obviously cannot be true under any interpretation, hence the origin of a closed tableau cannot be true under any interpretation—i.e. the origin of any closed tableau must be unsatisfiable. From this it follows that every formula provable by the tableau method must be a tautology. It therefore further follows that the tableau method is *consistent* in the sense that no formula and its negation are both provable (since no formula and its negation can both be tautologies).

*Completeness.* We now consider the more delicate converse situation: Is every tautology provable by the method of tableaux? Stated otherwise, if $X$ is a tautology, can we be sure that there exists at least one *closed* tableau starting with $FX$? We might indeed ask the following bolder question: If $X$ is a tautology, then will *every* complete tableau for $FX$

close? An affirmative answer to the second question would, of course, be even better than an affirmative answer to the first, since it would mean that any single completed tableau $\mathscr{T}$ for $F\,X$ would decide whether $X$ is a tautology or not.

Before the reader answers the question too hastily, we should consider the following. If we delete some of the rules for the construction of tableaux, it will *still* be true that a *closed* tableau for $F\,X$ *always* indicates that $X$ is a tautology. But if we delete too many of the rules, then we may not have left enough power to always derive a closed tableau for $F\,X$ whenever $X$ is a tautology. (For example, if we delete the first half of the conjunction rule, then it would be impossible to prove the tautology $(p \wedge q) \supset p$, though it would still be possible to prove $p \supset [q \supset (p \wedge q)]$. If we delete the second half but retain the first half, then we could prove the first tautology above, but not the second.) The question, therefore, is whether our present set of rules is *sufficient* to do this. Our present purpose is to show that they are sufficient.

We shall give the proof for tableaux using signed formulas (the modifications for tableaux using unsigned formulas are obvious—or indeed the result for tableaux for unsigned formulas follows directly from the result for tableaux with signed formulas.)

We are calling a branch $\theta$ of a tableau *complete* if for every $\alpha$ which occurs in $\theta$, both $\alpha_1$ and $\alpha_2$ occur in $\theta$, and for every $\beta$ which occurs in $\theta$, at least one of $\beta_1, \beta_2$ occurs in $\theta$. We call a tableau $\mathscr{T}$ *completed* if every branch of $\mathscr{T}$ is either closed or complete. We wish to show that if $\mathscr{T}$ is any *completed* open tableau (open in the sense that at least one branch is not closed), then the origin of $\mathscr{T}$ is satisfiable. More generally, we shall show

**Theorem 1.** *Any complete open branch of any tableau is (simultaneously) satisfiable.*

We shall actually prove something stronger. Suppose $\theta$ is a *complete open* branch of a tableau $\mathscr{T}$; let $S$ be the *set* of terms of $\theta$. Then the set $S$ satisfies the following three conditions (for every $\alpha$, $\beta$):

$H_0$: No signed *variable* and its conjugate are both in $S^1$).

$H_1$: If $\alpha \in S$, then $\alpha_1 \in S$ and $\alpha_2 \in S$.

$H_2$: If $\beta \in S$, then $\beta_1 \in S$ or $\beta_2 \in S$.

Sets $S$—whether finite or infinite—obeying conditions $H_0, H_1, H_2$ are of fundamental importance—we shall call them *Hintikka sets* (after Hintikka who studied their properties explicitly). We shall also refer to

---

[1]) Indeed no signed formula and its conjugate appear in $S$, but we do not need to involve this stronger fact.

Hintikka sets as sets which are *saturated downwards*. We shall also call any finite or denumerable *sequence* $\theta$ a *Hintikka sequence* if its set of terms is a Hintikka set.

Let us pause for a moment to compare the notion of *downward* saturation with that of saturation discussed earlier (cf. the preceding section on truth sets re-visited). The definition of a saturated set differs from that of a Hintikka set in that in $H_1, H_2$ "if" is replaced by "if and only if", and $H_0$ is strengthened to condition (0). So every saturated set is obviously also a Hintikka set. But a Hintikka set need not be saturated (e.g. any set of signed *variables* which contains no signed variable and its conjugate vacuously satisfies $H_1, H_2$, but such a set is certainly not saturated.)

Theorem 1 is substantially to the effect that every *finite* Hintikka set $S$ is satisfiable. The finiteness of $S$, however, is not needed in the proof (nor does it even simplify the proof), so we shall prove

**Hintikka's Lemma.** *Every downward saturated set $S$ (whether finite or infinite) is satisfiable.*

We remark that Hintikka's lemma is equivalent to the statement that every Hintikka set can be extended to a (i.e. is a subset of some) *saturated* set. We remark that Hintikka's lemma also holds for sets of *unsigned* formulas (where by a Hintikka set of unsigned formulas we mean a set $S$ satisfying $H_1, H_2$ and in place of $H_0$, the condition that no variable and its negation are both elements of $S$).

**Proof of Hintikka's Lemma.** Let $S$ be a Hintikka set. We wish to find an interpretation in which every element of $S$ is true. Well, we assign to each variable $p$, which occurs in at least one element of $S$, a truth value as follows:

(1) If $Tp \in S$, give $p$ the value true.

(2) If $Fp \in S$, give $p$ the value false.

(3) If neither $Tp$ nor $Fp$ is an element of $S$, then give $p$ the value true or false at will (for definiteness, let us suppose we give it the value true.)

We note that the directions (1), (2) are compatible, since no $Tp$ and $Fp$ both occur in $S$ (by hypothesis $H_0$). We now show that every element of $S$ is true under this interpretation. We do this by induction on the degree of the elements.

It is immediate that every signed *variable* which is an element of $S$ is true under this interpretation (the interpretation was constructed to insure just this). Now consider an element $X$ of $S$ of degree greater than 0, and suppose all elements of $S$ of lower degree than $X$ are true. We wish to show that $X$ must be true. Well, since $X$ is of degree greater than zero, it must be either some $\alpha$ or some $\beta$.

*Case 1.* Suppose it is an $\alpha$. Then $\alpha_1, \alpha_2$ must also be in $S$ (by $H_1$). But $\alpha_1, \alpha_2$ are of lower degree than $\alpha$. Hence by inductive hypothesis $\alpha_1$ and $\alpha_2$ are both true. This implies that $\alpha$ must be true.

*Case 2.* Suppose $X$ is some $\beta$. Then at least one of $\beta_1, \beta_2$ is in $S$ (by $H_2$). Whichever one is in $S$, being of lower degree than $\beta$, must be true (by inductive hypothesis). Hence $\beta$ must be true. This concludes the proof.

**Remark.** If we hadn't used the unifying "$\alpha, \beta$" notation, we would have had to analyze eight cases rather than two.

Having proved Hintikka's lemma, we have, of course, also proved Theorem 1. This in turn implies

**Theorem 2.** *(Completeness Theorem for Tableaux)*
(a) *If $X$ is a tautology, then every completed tableau starting with $F X$ must close.*
(b) *Every tautology is provable by the tableau method.*

To derive statement (a) from Theorem 1, suppose $\mathcal{T}$ is a complete tableau starting with $F X$. If $\mathcal{T}$ is open, then $F X$ is satisfiable (by Theorem 1), hence $X$ cannot be a tautology. Hence if $X$ is a tautology then $\mathcal{T}$ must be closed.

Let us note that for $S$ a finite Hintikka set, the proof of Hintikka's lemma effectively gives us an interpretation which satisfies $S$. Therefore, if $X$ is not a tautology, then a completed tableau for $F X$ provides us with a counterexample to $X$ (i. e. an interpretation in which $X$ is false).

**Example.** Let $X$ be the formula $(p \vee q) \supset (p \wedge q)$. Let us construct a completed tableau for $F X$:

$$
\begin{array}{ll}
& (1)\ F(p \vee q) \supset (p \wedge q) \\
& (2)\ T(p \vee q)\,(1) \\
& (3)\ F(p \wedge q)\,(1) \\
(4)\ Tp\,(2) & \qquad\qquad (5)\ Tq\,(2) \\
(6)\ Fp\,(3) \mid (7)\ Fq\,(3) & \quad (8)\ Fp\,(3) \mid (9)\ Fq\,(3) \\
X & \qquad\qquad\qquad\qquad X
\end{array}
$$

This tableau has two open branches. Let us consider the branch whose end point is (7). Acording to the method of Hintikka's proof, if we declare $p$ true and $q$ false, we have an interpretation which satisfies all lines of this branch. The reader can verify this by successively showing that (3), (2), (1) are true under this interpretation. Hence $F(p \vee q) \supset (p \wedge q)$ is true under this interpretation, which means $(p \vee q) \supset (p \wedge q)$ is false under this interpretation. Likewise the open branch terminating in (8) gives us another interpretation (viz. $q$ is true, $p$ is false) which is a counterexample to $X$.

*Tableaux for Finite Sets.* If $S$ is a *finite* set $\{X_1, ..., X_n\}$, by a tableau for $S$ is meant a tableau starting with

$$X_1$$
$$X_2$$
$$\vdots$$
$$X_n$$

and then continued using Rules $A$, $B$.

We leave it to the reader to modify our previous arguments for tableaux for single formulas, and prove:

**Theorem.** *A finite set $S$ is unsatisfiable iff there exists a closed tableau for $S$.*

We shall consider tableaux for *infinite* sets in a subsequent chapter.

*Exercise.* There is another way of proving the completeness theorem which does not use Hintikka's lemma.

Show (without use of the completeness theorem): (1) If there exists a closed tableau for $S \cup \{\alpha_1, \alpha_2\}$ then there exists a closed tableau for $S \cup \{\alpha\}$; (2) if there exist closed tableaux for $S \cup \{\beta_1\}$ and for $S \cup \{\beta_2\}$, then there exists a closed tableau for $S \cup \{\beta\}$; (3) if all elements of $S$ are of degree 0, and $S$ is unsatisfiable, then there exists a closed tableau for $S$ (trivial!).

Now define the degree of a finite set $S$ to be the sum of the degrees of the elements of $S$. Using (1), (2), (3) above, show by induction on the degree of $S$ that if $S$ is unsatisfiable, then there exists a closed tableau for $S$.

*Atomically Closed Tableaux.* Let us call a tableau *atomically* closed if every branch contains some *atomic* element and its *conjugate*. [By an *atomic* element we mean a propositional variable, if we are working with unsigned formulas, and a *signed* propositional variable if we are working with signed formulas. If we are working with unsigned formulas, then by an atomically closed tableau we mean a tableau in which every branch contains some propositional variable and its negation.]

Suppose we construct a completed tableau $\mathcal{T}$ for a set $S$, and declare a branch "closed" only if it is atomically closed. Now suppose $\mathcal{T}$ contains an (atomically) open branch $B$. Then the set of elements of $B$ is still a Hintikka set (because condition $H_0$ requires only that the set contain no *atomic* elements and its conjugate), hence is satisfiable (by Hintikka's lemma). We thus have:

**Theorem.** *If $S$ is unsatisfiable, then there exists an atomically closed tableau for $S$.*

**Corollary.** *If there exists a closed tableau for $S$, then there exists an atomically closed tableau for $S$.*

We remark that the above corollary can be easily proved directly (i.e. without appeal to any completeness theorem)—cf. Ex. 1 below.

*Exercise 1.* (*a*) Show directly by induction on the degree of $X$ that there exists an atomically closed tableau for any set $S$ which contains both $X$ and its conjugate $\bar{X}$.

(*b*) Using (*a*), show that any closed tableau can be further extended to an atomically closed tableau.

*Exercise 2.* Tableaux also provide a method of putting (unsigned) formulas into *disjunctive normal form* (cf. Ex. 3, end of Chapter I).

Suppose $\mathscr{T}$ is an open but completed tableau for (the unsigned formula) $X$. Let $B_1, \ldots, B_n$ be the open branches of $\mathscr{T}$, and for each $i \leqslant n$, let $C_i$ be the conjunction of the variables which appear in $B_i$ and those negations $\sim p$ (of variables) which appear in $B_i$ (the order of the conjunction is immaterial). Show that $C_1 \vee \cdots \vee C_n$ is a disjunctive normal form for $X$.

*Exercise 3.* Suppose we take Sheffer's stroke symbol $|$ as our sole connective for propositional logic (cf. Ex. 5 end of Chapter I). Show that the following tableau rules give a complete and correct system for propositional logic.

$$\text{Rule 1:} \quad \frac{F\,X\,|\,Y}{\begin{array}{c} TX \\ TY \end{array}} \qquad\qquad \text{Rule 2:} \quad \frac{T\,X\,|\,Y}{F\,X\,|\,F\,Y}$$

What are the corresponding two rules if we use $\downarrow$ instead of $|$ (cf. Exercise 5 at end of Chapter I)?

## Chapter III

# Compactness

### § 1. Analytic Proofs of the Compactness Theorem

Let us consider a finite or denumerable set $S$ of formulas. We recall that a Boolean valuation $v$ is said to *satisfy* $S$ iff all the elements of $S$ are true under $v$, and that $S$ is *satisfiable* iff $S$ is satisfied by at least one Boolean valuation.

We now consider the following question. Suppose that $S$ is a denumerably infinite set such that every finite subset of $S$ is satisfiable. Does it necessarily follow that $S$ is satisfiable? In other words, if for every finite subset $S_0$ of $S$ there is a Boolean valuation which satisfies $S_0$, does it follow that there is a Boolean valuation under which *all* the elements of $S$ are simultaneously true?

We can pose the question in another way. Consider $S$ arranged in some denumerable sequence $X_1, X_2, ..., X_n, ...$. To say that every finite subset of $S$ is satisfiable is to say nothing more nor less than that for each $n$, the set $\{X_1, ..., X_n\}$ is satisfiable. For clearly, if all finite subsets of $S$ are satisfiable, then for any $n$, the finite set $\{X_1, ..., X_n\}$ is satisfiable. Conversely, suppose that for each $n$, the set $\{X_1, ..., X_n\}$ is satisfiable. Then any finite subset $S_0$ of $S$ is a subset of $\{X_1, ..., X_n\}$ for some $n$, and hence is satisfiable.

We can thus look at the question as follows. Suppose that there is some Boolean valuation $v_1$ in which $X_1$ is true, and that there is a Boolean valutation $v_2$ (but not necessarily the same as $v_1$!) in which $X_1$ and $X_2$ are both true, and for each $n$ there is a Boolean valuation $v_n$ in which the first $n$ terms are true. Does there necessarily exist one Boolean valuation $v$ in which *all* the $X_i$ are simultaneously true?

We shall call a set $S$ *consistent* if every *finite* subset of $S$ is satisfiable (this is equivalent to saying that no formal contradiction can be derived from $S$ by the tableau method, i.e. there exists no finite number of elements $X_1, ..., X_n$ such that there is a closed tableau for $T(X_1 \wedge X_2 \wedge \cdots \wedge X_n)$). So the compactness question rephrased is whether a consistent infinite set is necessarily satisfiable—in other words, if it is impossible to derive a formal contradiction from $S$, is there necessarily an interpretation in which every element of $S$ is true? We shall return to this problem shortly.

**König's Lemma.** We first wish to consider a related problem, not about formulas of propositional logic, but about trees. Suppose $\mathcal{T}$ is a tree in which every branch is finite. Does it necessarily follow that $\mathcal{T}$ contains a branch of maximal length? Stated otherwise, if there exists no finite branch of maximal length, must $\mathcal{T}$ necessarily contain an infinite branch? Stated yet another way, if for every finite $n$, there is at least one point of level $n$, does $\mathcal{T}$ necessarily contain an infinite branch?

The answer for trees in general is "no", but for *finitely generated* trees (i.e. for trees in which each point has only finitely many successors) the answer is "yes".

A simple example of an infinitely generated tree for which the answer is "no" is the following:

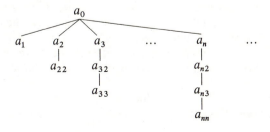

In this tree, $a_0$ branches to infinitely many points $a_1, a_2, ..., a_n, ...$ Every branch of this tree is finite, yet for each positive integer $n$, there is a branch of length greater than $n$ (the branch going through $a_{n+1}$).

Now König's lemma is to the effect that for a *finitely* generated tree, if for each $n$ there is at least one point of level $n$, then the tree must contain at least one infinite branch. Let us first observe that for a finitely generated tree $\mathcal{T}$, the statement that for every $n$ there is at least one point of level $n$ is equivalent to the statement that $\mathcal{T}$ has infinitely many points. (The first statement obviously implies the second, and conversely if $\mathcal{T}$ has infinitely many points, they must be scattered at infinitely many levels, since each level of a finitely generated tree contains only finitely many points.) We thus shall prove König's lemma in the following equivalent form.

**König's Lemma.** *Every finitely generated tree $\mathcal{T}$ with infinitely many points must contain at least one infinite branch.*

Many proofs of König's lemma exist in the literature. We shall give König's original proof, which is perhaps the shortest.

**Proof.** Call a point of $\mathcal{T}$ *good* if it has infinitely many descendents (i.e. if it dominates infinitely many points) and *bad* if it has only finitely many descendents. By hypothesis there are infinitely many points on $\mathcal{T}$, and they are all dominated by the origin; hence the origin is good.

We next observe that if all successors of a point are bad, then the point must be bad (since by hypothesis it has only finitely many successors). Thus a good point must have at least one good successor. Thus the origin $a_0$ has a good successor $a_1$, which in turn has a good successor $a_2$, which in turn has a good successor $a_3$, etc. In this way we generate an infinite branch $(a_0, a_1, a_2, ..., a_n ...)$.

**Remarks.** (1) For *unordered* trees, the axiom of choice is needed in the above proof, since at each stage we must *choose* a good successor. For *ordered* trees, the axiom of choice is not needed, since at each stage we can always choose the *leftmost* good successor.

(2) The crucial place where we used the hypothesis that $\mathcal{T}$ is finitely generated was in the statement that a good point necessarily has a good successor. This statement fails in general for infinitely generated trees— e. g. in the counterexample we considered a while back, the origin $a_0$ is good, but each of its successors $a_1, a_2, ..., a_n, ...$ is bad.

*The Compactness Problem Resumed.* The answer to the compactness question is "yes"—i.e. if all finite subsets of $S$ are satisfiable, then $S$ is satisfiable.

This theorem will be basic for our subsequent study of first order logic, and it will be profitable to consider several proofs of this result.

Roughly speaking, the proofs fall into two main lines. Suppose all finite subsets of $S$ are satisfiable. The proofs along the first line show that $S$ can then be extended to a Hintikka set, which in turn (by Hintikka's lemma) can be extended to a truth set. In the second type of proof (along the lines of Lindenbaum), we extend $S$ directly to a truth set[1]).

We now consider a proof along the first line—in this proof we use analytic tableaux and König's lemma.

Let $S$ be arranged in some denumerable sequence $X_1, X_2, ..., X_n, ...$ and suppose that for each $n$, the set $\{X_1, ..., X_n\}$ is satisfiable. Run a complete analytic tableau for $X_1$. This tableau cannot close since $X_1$ is satisfiable. Now "tack on" $X_2$ to the end of every open branch, and continue the tableau to completion. Again we must have at least one open branch, for otherwise $\{X_1, X_2\}$ would not be satisfiable. So tack on $X_3$ to the end of each open branch, and continue this process indefinitely (successively tacking on $X_4, X_5$, etc.). At no stage can the tableau close, for otherwise for some finite $n$, the set $\{X_1, ..., X_n\}$ would not be satisfiable. We thus obtain an *infinite* tree. Then by König's lemma, the tableau contains at least one infinite branch $\theta$, which clearly must be open. It is obvious that $\theta$ contains all the $X_i$, and the set of terms of $\theta$ is easily seen to be a Hintikka set. Then by Hintikka's lemma, this set is satisfiable, hence also the subset $S$. This concludes the proof.

**Discussion.** The above construction has shown something about tableaux which is of interest on its own account. We have previously defined a tableau for a single formula $X$. Let us now define a tableau for a set $S$ (whether finite or infinite) as a tree constructed as follows. We begin by placing any element of $S$ at the origin. Then at any stage we may extend the tableau by Rule $A$ or Rule $B$ or we may add any element of $S$ to the end of any open branch. This completes the definition of a tableau *for S*. Call a point of a tableau *fulfilled* if it is either of degree 0, or it is an $\alpha$ and $\alpha_1$, $\alpha_2$ both appear on every open branch passing through the point, or it is a $\beta$ and every open branch passing through the point contains either $\beta_1$ or $\beta_2$. To say that a tableau is *complete* (as defined in Chapter II) is to say that every point is fulfilled. Our proof above provided a scheme for constructing a *complete* tableau for any denumerably infinite set $S$.

Many such schemes could be given. Here is another one (which incidentally comes closer to something we shall do in First Order Logic). Start the tableau with $X_1$. Then fulfill the origin, and then append $X_2$ to every open branch. This concludes the first stage. Next fulfill all

---

[1]) The 2 proofs generalize to distinct theorems in a more abstract setting. We discuss this and related topics in Smullyan, [4].

points of level 2, then append $X_3$ to every open branch. This concludes the second stage. Then fulfill all points of level 3, append $X_4$ to all open branches, etc. Obviously every point of this tableau gets fulfilled.

We now wish to point out that although we used both analytic tableaux and König's lemma in our proof of the compactness theorem, neither is really essential, as the following construction will show.

We employ the following notation. For any set $S$ and formula $X$, by $\{S, X\}$ we mean $S \cup \{X\}$—i.e. the set whose elements are $X$ and those of $S$. By $\{S, X, Y\}$ we mean $S \cup \{X, Y\}$—i.e. the set whose elements are $X, Y$ and those of $S$. More generally we write $\{S, X_1, ..., X_n\}$ to mean $S \cup \{X_1, ..., X_n\}$.

We have defined a set to be consistent if all its finite subsets are satisfiable. We say that $S$ is *inconsistent* if $S$ is not consistent—equivalently, if some finite subset of $S$ is unsatisfiable. Obviously any extension (superset) of an inconsistent set is inconsistent; any subset of a consistent set is consistent. It is almost immediate that consistency satisfies the following conditions (for any set $S$ and any $\alpha, \beta$).

$C_0$: No set containing a propositional variable and its negation is consistent.

$C_1$: If $\{S, \alpha\}$ is consistent, so is $\{S, \alpha_1, \alpha_2\}$.

$C_2$: If $\{S, \beta\}$ is consistent, then *at least one* of the sets $\{S, \beta_1\}$ or $\{S, \beta_2\}$ is consistent.

We can equivalently state $C_0, C_1, C_2$ in the following forms.

$I_0$: Any set containing a propositional variable and its negation is inconsistent (i.e. not consistent).

$I_1$: If $\{S, \alpha_1, \alpha_2\}$ is inconsistent, so is $\{S, \alpha\}$.

$I_2$: If $\{S, \beta_1\}$ and $\{S, \beta_2\}$ are both inconsistent, so is $\{S, \beta\}$.

$I_0$ is immediate, since for any propositional variable $p$, the set $\{S, p, \sim p\}$ has the unsatisfiable finite subset $\{p, \sim p\}$.

As for $I_1$, suppose $\{S, \alpha_1, \alpha_2\}$ is inconsistent. Then there must be a finite subset $S_1$ of $S$ such that $\{S_1, \alpha_1, \alpha_2\}$ is unsatisfiable (because there is a finite subset $S_0$ of $\{S, \alpha_1, \alpha_2\}$ which is unsatisfiable, hence $\{S_0, \alpha_1, \alpha_2\}$ is unsatisfiable, and we can take $S_1$ to be the elements of $S_0$ other than $\alpha_1, \alpha_2$, so that $\{S_1, \alpha_1, \alpha_2\}$ is the same as $\{S_0, \alpha_1, \alpha_2\}$ but $S_1$ is a subset of $S$). Then $\{S_1, \alpha\}$ is unsatisfiable (because in any interpretation in which $\alpha$ is true, $\alpha_1, \alpha_2$ must also be true). Furthermore $\{S_1, \alpha\}$ is a finite subset of $\{S, \alpha\}$ so $\{S, \alpha\}$ is inconsistent.

As for $I_2$, suppose $\{S, \beta_1\}$ and $\{S, \beta_2\}$ are both inconsistent. Then there must be finite subsets $S_1, S_2$ of $S$ such that $\{S_1, \beta_1\}$ and $\{S_2, \beta_2\}$ are both unsatisfiable. (Cf. second statement in the above proof of $I_1$.) Let $S_3$ be the union $S_1 \cup S_2$. Then $\{S_3, \beta_1\}$ and $\{S_3, \beta_2\}$ are both unsatisfiable. Hence $\{S_3, \beta\}$ is unsatisfiable (why?), but $S_3$ is a finite subset of $S$, so $\{S_3, \beta\}$ is a finite subset of $\{S, \beta\}$. Therefore $\{S, \beta\}$ is inconsistent.

**Remark.** Of course $C_0$, $C_1$, $C_2$ are immediate consequences of the compactness theorem, since the compactness theorem implies that consistency is the same as satisfiability (and obviously $C_0$, $C_1$, $C_2$ hold replacing "consistent" by "satisfiable"). But we wish to use $C_0$, $C_1$, $C_2$ to give an alternative proof of the compactness theorem.

We remark that conditions $C_0$, $C_1$, $C_2$ are the only properties of consistency which we shall need for our proof. That is to say, given any other definition of "consistency" satisfying $C_0$, $C_1$, $C_2$, it will still hold that any $S$ which has this "consistency" property is satisfiable. (This fact will be of particular importance later on when we discuss an abstract treatment of consistency.)

Now for the construction. Let $S$ be a consistent set arranged in some denumerable sequence $X_1, X_2, \ldots, X_n, \ldots$. We wish to generate an infinite Hintikka sequence $\theta$ whose terms include each $X_i$. At each stage of our construction, we will have generated a finite number $k \geqslant n$ of terms $Y_1, \ldots, Y_k$ such that $S \cup \{Y_1, \ldots, Y_k\}$ is consistent, and our next act will consist of adding 1, 2, or 3 more terms in such a manner that the union of the enlarged set of terms with $S$ is still consistent. Specifically, our generation process is given by the following inductive scheme.

We start the sequence by taking $X_1$ as the first term. This concludes the first stage.

Now suppose that we have completed the $n$th stage, and have at hand a finite sequence $X_1, Y_2, \ldots, Y_{n+i}$ $(i \geqslant 0)$, which we will call $\theta_n$, such that $\theta_n$ is consistent with $S$ (in the sense that $\{S, Y_1, Y_2, \ldots, Y_{n+i}\}$ is consistent). Now we look at the $n$th term $Y_n$, and our next act is as follows.

(a) If $Y_n$ is an $\alpha$, then we take $\theta_{n+1}$ to be the sequence $(Y_1, Y_2, \ldots, Y_{n+i}, \alpha_1, \alpha_2, X_{n+1})$ (in other words we adjoin both $\alpha_1$ and $\alpha_2$ and *also* add the element $X_{n+1}$ from our set $S$). It follows readily from condition $C_1$ that $\theta_{n+1}$ is again consistent with $S$.

(b) If $Y_n$ is some $\beta$ then it follows from $C_2$ that either $(\theta_n, \beta_1, X_{n+1})$ (i.e. $(Y_1, Y_2, \ldots, Y_{n+i}, \beta_1, X_{n+1})$) is consistent with $S$ or $(\theta_n, \beta_2, X_{n+1})$ is consistent with $S$. We let $\theta_{n+1}$ be the former, if consistent with $S$, or the latter otherwise.

(c) If $Y_n$ is neither an $\alpha$ nor a $\beta$ (i.e. if it is either a propositional variable or its negation), then we merely adjoin $X_{n+1}$ to $\theta_n$ (i.e. $\theta_{n+1} = (Y_1, Y_2, \ldots, Y_{n+i}, X_{n+1})$).

This concludes the scheme of our construction. Each $X_n$ is introduced during the $n$th stage, so every $X_i$ appears in $\theta$. It is obvious that conditions $H_1$, $H_2$ of a Hintikka set are fulfilled (for each $\alpha$ which appears as some $n$th term of $\theta$, $\alpha_1$, $\alpha_2$ are both added to $\theta$ at stage $n+1$, and for each $\beta$ which appears as some $n$th term, either $\beta_1$ or $\beta_2$ is added at stage $n+1$). It remains to verify condition $H_0$. Well, since each $\theta_n$ is consistent with $S$, then it follows from $C_0$ that no propositional variable and its negation both occur in $\theta_n$. Hence no propositional variable and its negation both appear in $\theta$. This concludes the proof.

**Discussion.** The above proof nowhere uses König's lemma, nor does it explicitly appeal to analytic tableaux. However the above construction in effect generates the *leftmost* infinite branch of the complete tableau for $S$ obtained by the second of our earlier constructions. The point is that we can generate this branch without considering the other branches at all.

*Exercise.* Suppose $S$ is a set such that for any interpretation $v_0$, there is always at least one element of $S$ which is true under $v_0$. Show that there is a finite subset $\{X_1, ..., X_n\}$ of $S$ such that the disjunction $(X_1 \vee \cdots \vee X_n)$ is a tautology. (Hint: This can be shown as an almost immediate consequence of the compactness theorem. It is not necessary to duplicate (or rather "dualize") the construction we made in order to prove the compactness theorem.)

## § 2. Maximal Consistency: Lindenbaum's Construction

A set $M$ (of formulas) is called *maximally consistent* if it is consistent and if no proper extension of $M$ is consistent. (By a *proper* extension of a set $S$ we mean a superset of $S$ which contains at least one element not in $S$). Any truth set $W$ is obviously consistent (since it, and hence each of its subsets, is satisfiable). And every element outside a truth set is false, hence cannot be adjoined to $W$ without destroying consistency. So any truth set is maximally consistent. We now wish to show

**Lemma 1.** *Any maximally consistent set is a truth set.*

Our proof uses the following properties of consistency (which indeed are the *only* properties of consistency which we need):

$L_0$: If $S$ is consistent, then every finite subset of $S$ is satisfiable.

$L_1$: If $S$ is consistent, then for any formula $X$, at least one of the two sets $\{S, X\}$, $\{S, \sim X\}$ is consistent.

$L_0$ is immediate from the definition of consistency. As for $L_1$, suppose that $\{S, X\}$ and $\{S, \sim X\}$ are both inconsistent. Then there is a finite subset $S_1$ of $S$ such that $\{S_1, X\}$ is unsatisfiable, and there is a finite subset $S_2$ of $S$ such that $\{S_2, \sim X\}$ is unsatisfiable. Let $S_3 = S_1 \cup S_2$. Then $\{S_3, X\}$ and $\{S_3, \sim X\}$ are both unsatisfiable, hence $S_3$ is unsatisfiable, but $S_3$ is a finite subset of $S$. Therefore $S$ is inconsistent.

$L_1$ at once implies

$L_1^*$: If $M$ is maximally consistent, then for any $X$ either $X \in M$ or $\sim X \in M$.

Now we turn to the

**Proof of Lemma 1.** Let $M$ be maximally consistent. Since $M$ is consistent, then for any formula $X$, at least one of $X$, $\sim X$ lies outside $M$ (by $L_0$). But by $L_1^*$, at least one of $X$, $\sim X$ lies in $M$. Thus our first condition (0) of truth sets is satisfied (cf. Chapter II, the section called "truth sets revisited"). It remains to show that for any $\alpha$, $\alpha$ belongs to $S$ iff $\alpha_1$, $\alpha_2$ both belong to $S$ (for as we pointed out in Chapter II, it then follows from (0), (a) that $\beta$ is in $S$ iff at least one of $\beta_1$, $\beta_2$ is in $S$). Well, suppose $\alpha \in M$. Then $(\sim \alpha_1)$ is not in $M$ (by $L_0$, since $\{\alpha, \sim \alpha_1\}$ is not satisfiable). Hence $\alpha_1 \in M$ (by $L_1^*$). Similarly $\alpha_2 \in M$. Conversely suppose $\alpha_1 \in M$, $\alpha_2 \in M$. Then $(\sim \alpha) \notin M$ (by $L_0$, since $\{\alpha_1, \alpha_2, \sim \alpha\}$ is not satisfiable). Hence $\alpha \in M$ (by $L_1^*$). This concludes the proof.

**Lindenbaum's Theorem.** *Every consistent set can be extended to a maximally consistent set.*

Before discussing Lindenbaum's proof, we note that the following properties of consistency are trivial.

$F_1$: A finite set is consistent iff it is satisfiable.

$F_2$: A (possibly finite) set is consistent iff all its finite subsets are consistent.

$F_1$ holds, because if a finite set is consistent, then (by definition) all its finite subsets, including itself, are satisfiable. Conversely, if all finite subsets of a finite set $S$ are satisfiable, then by definition $S$ is consistent. $F_2$ then follows immediately from $F_1$.

A property $P$ of sets is said to be of *finite character* if for any set $S$, it has the property $P$ iff all finite subsets of $S$ have the property $P$. Condition $F_2$ thus says that consistency is a property of finite character. The fact that consistency is of finite character is actually the only property of consistency used in Lindenbaum's proof. Indeed Lindenbaum's theorem can be looked at as a special case of Tukey's lemma, which we now discuss.

Consider an arbitrary universe (set) $U$ of objects (maybe finite, denumerable or non-denumerable). By "set" let us temporarily understand "subset of $U$". Consider now a property $P$ of these sets. By a maximal set having property $P$ is meant a set $S$ having the property but such that no proper extension of $S$ (still within $U$) has the property. Tukey's lemma says that if $P$ is of *finite character* then any set having property $P$ can be extended to a *maximal* set having property $P$. This lemma (for arbitrary universe $U$) is well known to be equivalent to the Axiom of Choice. Of course, for a *denumerable* universe $U$, the Axiom of Choice is not needed. We shall therefore use what is essentially Lindenbaum's argument and show

**Theorem.** *( Tukey's lemma for the denumerable case ). For any denumerable universe $U$ and any property $P$ of subsets of $U$ of finite character any set $S$ (of elements of $U$) having property $P$ can be extended to a maximal subset of $U$ having property $P$.*

**Proof.** (After Lindenbaum). We first arrange all elements of $U$ (not just those of $S$) in some denumerable sequence $Y_1, Y_2, \ldots, Y_n, \ldots$. Now we generate a denumerable sequence $S_0, S_1, S_2, \ldots, S_n, \ldots$ of sets by the following inductive scheme. We set $S_0 = S$. Now assume $S_n$ has been defined. We then determine $S_{n+1}$ as follows. If $S_n \cup \{Y_{n+1}\}$ has property $P$, we let $S_{n+1} = S_n \cup \{Y_{n+1}\}$. If not, then we let $S_{n+1} = S_n$.

It is obvious that we have $S_0 \subseteq S_1 \subseteq S_2 \subseteq \cdots \subseteq S_n \subseteq S_{n+1} \subseteq \ldots$, and that each $S_i$ has property $P$. We now let $M$ be the *union* of all the sets $S_i$—i.e. an element $X$ of $U$ belongs to $M$ iff $X$ belongs to at least one $S_i$. Obviously $M$ is an extension of $S_0$, and we assert that $M$ is a *maximal* set having property $P$.

First we must show that $M$ has property $P$. Well, let $K$ be any finite subset of $M$. Then $K$, being finite, must be a subset of some $S_i$ (why?). Since $S_i$ has property $P$ and $P$ is of finite character, then $K$ has property $P$. Thus every finite subset $K$ of $M$ has property $P$, therefore $M$ has property $P$ (since $P$ is of finite character).

As to maximality, take any $Y_i$ such that $M \cup \{Y_i\}$ has property $P$. We must show that $Y_i$ lies in $M$. Since $M \cup \{Y_i\}$ has property $P$, so does the subset $S_i \cup \{Y_i\}$ (for any property $P$ of finite character, if a set $S$ has the property, so does any subset $S'$, since all finite subsets of $S'$ are also subsets of $S$). Then $Y_i \in S_{i+1}$, so $Y_i$ belongs to $M$. This concludes the proof.

*Exercise.* Call a set $S$ *complete* if every formula or its negation is in $S$. We have already shown that every maximally consistent set is complete. Show that any consistent complete set must be maximally consistent (and thus a consistent set is maximally consistent if and only if it is complete).

### § 3. An Analytic Modification of Lindenbaum's Proof

We proved the compactness theorem in § 1 by showing how to extend a consistent set $S$ to a Hintikka set. Furthermore, the elements of the Hintikka set of the construction were all subformulas of elements of $S$—or at worst *negations* of such subformulas. In this sense, we say that the methods of § 1 are *analytic*. (We use the term "analytic" in a way which rather corresponds to the use of the phrase "cut free" in

Gentzen's work; we will subsequently study the relationship quite closely.) By contrast, the maximally consistent set $M$ obtained by Lindenbaum's construction contains *every* formula $X$ or its negation (even if neither is a subformula of some element of $S$).

Lindenbaum's construction is rather simpler than the construction of § 1, which involves a rather careful scheme for generating the infinite Hintikka branch $\theta$. We now wish to show how Lindenbaum's construction can be modified along analytic lines (which will have applications further on).

We must first get out of the way a certain detail about subformulas and their negations. Define $Y$ to be a *direct descendant* of $X$ if either $X$ is some $\alpha$ and $Y$ is $\alpha_1$ or $\alpha_2$, or $X$ is some $\beta$ and $Y$ is $\beta_1$ or $\beta_2$. And define $Y$ to be a *descendant* of $X$ if there exists a finite sequence beginning with $X$ and ending with $Y$ such that each term of the sequence (other than the first) is a direct descendant of the preceding term. It is easily verified that if $Y$ is a descendant of $X$, then $Y$ is either a subformula of $X$ or the negation of a subformula of $X$ (though it is not in general the case that the negation of a subformula of $X$ is a descendant of $X$—e.g. $\sim X$ is the negation of a subformula of $X \wedge Y$, but it is not a descendant of $X \wedge Y$). We shall define $S^0$ to be the set of all descendants of elements of $S$. Our earlier remark that the constructions of § 1 use only subformulas of elements of $S$ and their negations can really be sharpened to the fact that the constructions use only elements of $S^0$.

Now suppose $S$ is consistent. Since consistency is of finite character, we can extend $S$ to a maximally consistent *subset of* $S^0$—i.e. to a consistent subset of $S^0$ such that no proper extension contained within $S^0$ is consistent. Now this maximally consistent subset of $S^0$ is in general not a truth set. But we show

**Theorem.** *Every maximally consistent subset of $S^0$ is a Hintikka set.*
    The proof follows directly from the properties $C_0, C_1, C_2$ of consistency which we used in § 1. Let $M$ be a maximally consistent subset of $S^0$. Since $M$ is consistent, then $M$ contains no variable and its negation (by $C_0$). This proves $H_0$. As to $H_1$, suppose $\alpha \in M$. Then $M \cup \{\alpha\} = M$, so $M \cup \{\alpha\}$ is consistent. Then, by $C_1, M \cup \{\alpha_1\}$ is consistent, hence $\alpha_1 \in M$ (by maximality, since $\alpha_1 \in S^0$). Likewise $\alpha_2 \in M$. As to $H_2$, suppose $\beta \in M$. Then $M = M \cup \{\beta\}$, and $M \cup \{\beta\}$ is consistent. Then by $C_2$, either $M \cup \{\beta_1\}$ or $M \cup \{\beta_2\}$ is consistent, so $\beta_1$ (respectively $\beta_2$) belongs to $M$ (again since $M$ is a maximal consistent subset of $S^0$ and $\beta_1, \beta_2$ certainly belong to $S^0$). This completes the proof.

*Exercise.* There is another proof of the compactness theorem, which is well known and probably the simplest of all. (We did not emphasize it because the principles involved will not be as useful to us in other contexts as the principles of § 1, § 2). Arrange all the *propositional variables* in a denumerable sequence $p_1, p_2, ..., p_n, ....$ Define the sequence $B_1, B_2, ..., B_n$ of sets by the following inductive scheme. If $S \cup \{p_1\}$ is consistent, let $B_1 = \{p_1\}$, otherwise let $B_1 = \{\sim p_1\}$. Suppose $B_n$ is defined. Then let $B_{n+1} = B_n \cup \{p_{n+1}\}$ if $S \cup B_n \cup \{p_{n+1}\}$ is consistent, otherwise let $B_{n+1} = B_n \cup \{\sim p_{n+1}\}$. Then let $B$ be the union of $B_1, B_2, ..., B_n, ....$ Show that there is exactly one interpretation which satisfies $B$, and that all elements of $S$ are true under this interpretation (and hence that $S$ is satisfiable).

## § 4. The Compactness Theorem for Deducibility

We shall find the compactness theorem particularly useful in the following form. We shall say that $X$ is *deducible* from a set $S$ if there are finitely many elements $X_1, ..., X_n$ of $S$ such that the formula $(X_1 \wedge \cdots \wedge X_n) \supset X$ is a tautology. Equivalently, $X$ is deducible from $S$ if $\sim X$ is inconsistent with $S$ (or equivalently with some finite subset of $S$).

**Compactness Theorem.** *(Second Form). If $X$ is true in all Boolean valuations which satisfy $S$, then $X$ is deducible from $S$.*

**Proof.** By hypothesis $\{S, \sim X\}$ is unsatisfiable. Then by the compactness theorem, some finite subset $S_0$ of $\{S, \sim X\}$ is unsatisfiable. Letting $X_1, ..., X_n$ be those elements of $S_0$ other than $\sim X$ (if $\sim X$ happens to be in $S_0$), then $\{X_1, ..., X_n, \sim X\}$ is unsatisfiable, so $(X_1 \wedge \cdots \wedge X_n) \supset X$ is a tautology.

*Exercise.* A set $S$ is called deductively closed if every formula which is deducible from $S$ lies in $S$. Prove Tarski's theorem: A consistent deductively closed set is the intersection of all its complete consistent extensions (complete in the sense that every formula or its negation belongs to the set).

## Part II

# First-Order Logic

Chapter IV

# First-Order Logic. Preliminaries

## § 1. Formulas of Quantification Theory

For first order logic (quantification theory) we shall use the following symbols:

(*a*) The symbols of propositional logic other than propositional variables.

(*b*) $\forall$ [read "for all"],

$\exists$ [read "there exists"].

($c_1$) A denumerable list of symbols called *individual variables*.

($c_2$) A denumerable list of symbols (not in $c_1$) called *individual parameters*.

(*d*) For each positive integer *n*, a denumerable list of symbols called *n-ary predicates*, or *predicates of degree n*.

The symbols "$\forall$" and "$\exists$" are respectively called *universal* and *existential* quantifiers. The term "variable" shall henceforth mean *individual variable* (they should not be confused with the propositional variables used earlier). We shall use small letters "$x$", "$y$", "$z$" to denote arbitrary individual variables. We shall use "$a$", "$b$", with or without subscripts to denote individual parameters (henceforth called just "parameters"). We shall use capital letters "$P$", "$Q$", "$R$", with or without subscripts to denote predicates; their degree will always be clear from the context. We shall use the term "individual symbols" collectively for (individual) variables and parameters.

*Atomic Formulas.* By an atomic formula (of quantification theory) we mean an $(n+1)$-tuple $P c_1, ..., c_n$, where $P$ is any predicate of degree $n$ and $c_1, ..., c_n$ are any individual symbols (variables or parameters).

*Formulas.* Starting with the atomic formulas, we build the set of all formulas (of quantification theory) by the formation rules of propositional logic, together with the rule:

(1) If $A$ is a formula and $x$ is a variable, then both $(\forall x)A$ and $(\exists x)A$ are formulas; these are respectively called the *universal* quantification of $A$ with respect to $x$ and the *existential* quantification of $A$ with respect to $x$.

The above recursive definition of "formula" can be made explicit as follows: $A$ is a formula iff there is a finite sequence of expressions which terminates with $A$ and is such that each term is either an atomic formula or is the negation, conjunction, disjunction or conditional of 1 or 2 earlier terms, or is the existential or universal quantification of some earlier term with respect to some variable $x$. Such a sequence is again called a *formation* sequence for $A$.

*Pure Formulas.* By a pure formula, we mean one with no parameters.

*Degrees.* By the *degree* $d(A)$ of a formula $A$ we mean the number of occurrences of logical connectives and quantifiers. Thus

(1) Every atomic formula is of degree 0.

(2)     $d(\sim A) = d(A) + 1$

$d(A \wedge B) = d(A \vee B) = d(A \supset B) = d(A) + d(B) + 1$

(3) $d((\forall x)A) = d(A) + 1$

$d((\exists x)A) = d(A) + 1$

*Substitution.* For every formula $A$, variable $x$, and parameter $a$, we define the formula $A_a^x$ by the following inductive scheme:

(1) If $A$ is *atomic*, then $A_a^x$ is the result of substituting $a$ for every occurrence of $x$ in $A$.

(2)     $[A \wedge B]_a^x = A_a^x \wedge B_a^x$

$[A \vee B]_a^x = A_a^x \vee B_a^x$

$[A \supset B]_a^x = A_a^x \supset B_a^x$

$[\sim A]_a^x = \sim [A_a^x]$

(3) $[(\forall x)A]_n^x = (\forall x)A$

$[(\exists x)A]_a^x = (\exists x)A$

But for a variable $y$ distinct from $x$

$[(\forall x)A]_a^y = (\forall x)[A_a^y]$

$[(\exists x)A]_a^y = (\exists x)[A_a^y]$

*Example.* If $A$ is the formula $(\forall x)Px \vee \sim(\exists y)Qxy$, then $A_a^x = (\forall x)Px \vee \sim(\exists y)Q(a, y)$.

We refer to $A_a^x$ as the result of substituting $a$ for every free occurrence of $x$ in $A$ (cf. discussion below).

By a *closed* formula or a *sentence* we mean a formula $A$ such that for every variable $x$ and every parameter $a$, $A_a^x = A$. [This is equivalent to saying, in more usual terms, that no variable has a free occurrence in $A$ —cf. following discussion.]

**Discussion.** In the way we are treating quantification theory, it is not necessary to define the notions of "free occurrences" and "bound occurrences". But to make better contact with the more conventional treatments, some remarks are in order. An expression $(\forall x)$, or $(\exists x)$ we shall call a *quantified variable*. [It is more usually called a "quantifier", but we prefer the latter term for the symbols $\forall$, $\exists$ themselves.]

In any formula $A$, by the scope of an occurrence of a quantified variable, is meant the smallest formula which follows that occurrence.

*Examples.* Consider the following 3 formulas:

(a) $((\forall x)Px) \supset [(\forall x)Qxy \vee Rx]$

(b) $(\forall x)[Px \supset [(\forall x)Qxy \vee Rx]]$

(c) $(\forall x)[Px \supset (\forall x)(Qxy \wedge Rx)]$

In each of these, there are 2 occurrences of "$(\forall x)$"; let $0_1$ be the first (leftmost) one, and $0_2$ be the second occurrence. In (a), the scope of $0_1$ is $Px$. In (c) the scope of $0_1$ is $(Px \supset (\forall x)(Qxy \wedge Rx))$. The scope of $0_2$ in both (a) and (b) is $Qxy$; the scope of $0_2$ in (c) is $(Qxy \vee Rx)$.

We define an occurrence of a variable $x$ in a formula $A$ to be *bound* if it is either within the scope of some occurrence of $(\forall x)$ or $(\exists x)$, or else is itself immediately preceded by $\forall$ or $\exists$. An occurrence of $x$ in $A$ is called *free* if it is not bound. Finally $x$ is said to have a *free occurrence* in $A$ if at least one occurrence of $x$ in $A$ is free.

In the above examples, $x$ has no free occurrence in either (b) or (c), and has only one free occurrence in (a). All occurrences of $y$ in (a), (b), (c) are free.

The *semantical* significance of the notion of free and bound occurrences can be illustrated by the following example:

Consider the expression

(1) $$x = 5y.$$

The truth or falsity of (1) depends both on a choice of value for $x$ and a choice for $y$. This reflects the purely syntactical fact that $x$ and $y$ each have a free occurrence in (1).

Now consider:

(2) $$(\exists y)[x = 5y].$$

The truth or falsity of (2) depends on $x$, but not on any choice for $y$. Indeed, we could restate (2) in a form in which the variable $y$ does not even occur; viz. "$x$ is divisible by 5". And this reflects the fact that $x$ has a free occurrence in (2) but $y$ does not.

We remark that $A_a^x$ is indeed the result of substituting $a$ for every free occurrence of $x$ in $A$—this can be verified by induction on the degree of $A$.

We also remark that it is possible to define what it means for $x$ to have a free occurrence in $A$ without having to define the notion of "occurrence". Using our substitution operation, we can say that $x$ has a free occurrence in $A$ if for some parameter $a$ (or equivalently for *every* parameter $a$) $A_a^x$ is distinct from $A$. Without using the substitution operation, we can alternatively characterize the notion "$x$ has a free occurrence in $A$" by the following inductive rules:

(1) $x$ has a free occurrence in an atomic formula $Pc_1, \ldots, c_n$ iff $x$ is identical with one of the symbols $c_1, \ldots, c_n$.

(2) For each of the binary connectives $b$, $x$ has a free occurrence in $A\,b\,B$ iff $x$ has a free occurrence in at least one of $A, B$. And $x$ has a free occurrence in $\sim A$ iff $x$ has a free occurrence in $A$.

(3) If $y, x$ are distinct variables then $y$ has a free occurrence in $(\forall x)A$ [and likewise $(\exists x)A$] iff $y$ has a free occurrence in $A$. And $x$ does *not* have a free occurrence in $(\forall x)A$, nor in $(\exists x)A$.

*Subformulas and Formation Trees.* "Formula" will henceforth mean "closed formula", unless specified to the contrary. The notion of *immediate* subformula (in the sense of quantification theory) is given explicitly by the rules:

(1) $A, B$ are immediate subformulas of $A \wedge B$, $A \vee B$, $A \supset B$, and $A$ is an immediate subformula of $\sim A$.

(2) For any parameter $a$, variable $x$, formula $A$, $A_a^x$ is an immediate subformula of $(\forall x)A$ and of $(\exists x)A$.

The definition of "subformula" is like that of Chapter I, but construing "immediate subformula" in the above sense.

By a *formation* tree (in the sense of first order logic) is meant a tree in which each end point is atomic, and such that for every other point one of the following conditions holds:

(1) It is of the form $A\,b\,B$, and has $A$, $B$ for its 1st, 2nd successors, and it has no other successors, or it is of the form $\sim A$, and has $A$ for its sole successor.

(2) It is of the form $(\forall x)A$ or $(\exists x)A$ and it has $A_{a_1}^x$, $A_{a_2}^x$, ..., $A_{a_n}^x$, ... as its successors (say as the 1st, 2nd, ... $n$th ... respectively, though the order is not of any real significance).

Since we have denumerably many parameters, formation trees for quantification theory are not finitely generated (they involve infinite branching) unlike the case for propositional logic. [However, it is obvious that all branches are finite—indeed the length of each branch is bounded by $1 +$ the degree of the formula at the origin.]

Again we call $\mathcal{T}$ a formation tree *for* $A$ if $\mathcal{T}$ is a formation tree with $A$ at the origin. And again, the subformulas of $A$ are precisely the formulas which occur in at least one place of the formation tree for $A$.

## § 2. First-Order Valuations and Models

*Formulas with Constants in* $\cup$. We let $\cup$ be any non-empty set. We shall call $\cup$ a *universe of individuals*—more briefly, a *universe* or a *domain*. We now wish to define the notions of formulas *with constants* in $\cup$—or more briefly "$\cup$-formulas".

By an *atomic* $\cup$ -formula, we shall mean an ordered $(n+1)$-tuple $P\mathscr{E}_1, ..., \mathscr{E}_n$ (also written $P\mathscr{E}_1 ... \mathscr{E}_n$), where $P$ is an $n$-ary predicate, and each $\mathscr{E}_i$ is either a variable or an element of U. [Note that we do not allow any $\mathscr{E}_i$ to be a parameter.] Having defined the *atomic* U-formulas, we can then define the set of all U-formulas by the formation rules given in § 1. Thus a U-formula is like a formula with parameters, except that it contains elements of U in place of parameters. This includes the "pure" formulas—i.e. those with no parameters, and no constants in U—as special cases. For any element $k \in$ U we define the formula $F_k^x$ in exactly the same manner as we define $F_a^x$, where $a$ is a parameter. If $F$ is a U-formula, so is $F_k^x$. We also define the notion of a U-subformula, analogous to that of "subformula", and "formation trees" for U-formulas in again an analogous manner [however, we do not require any ordering of U; formation trees are not required to be ordered trees]. We let $E^U$ be the set of all *closed* U-formulas. By a *first order valuation* $v$ of $E^U$ we mean an assignment of truth values to all elements of $E^U$ such that for every $A$ in $E^U$ and every variable $x$, the following conditions hold:

$F_1$: $v$ is a *Boolean* valuation of $E^U$.

$F_2$: (a) $(\forall x)A$ is true under $v$ iff for every $k \in$ U, $A_k^x$ is true under $v$.

(b) $(\exists x)A$ is true under $v$ iff for at least one element $k \in$ U, $A_k^x$ is true under $v$.

By a *first order truth* set (with respect to the universe U) we mean a subset $S$ of $E^U$ which satisfies all the conditions of a truth set as defined in propositional logic, as well as the conditions:

(a) $(\forall x)A$ belongs to $S$ iff for every $k \in$ U, $A_k^x$ belongs to $S$.

(b) $(\exists x)A$ belongs to $S$ iff for at least one $k$ in U, $A_k^x$ belongs to $S$.

Again it is immediate that $S$ is a first order truth set iff the characteristic function of $S$ is a first order valuation. [By the characteristic function of $S$ we mean, of course, relative to $E^U$—i.e., the function which assigns $t$ to every element of $S$ and $f$ to every element of $E^U$ which is not in $S$.]

*Atomic Valuations.* By an *atomic* valuation of $E^U$ we mean an assignment of truth values to all *atomic* elements of $E^U$. If two first order valuations agree on all atomic elements of $E^U$, then they must agree on every element of $A$ of $E^U$ (this is easily seen by induction on the degree of $A$). Thus an atomic valuation $v_0$ of $E^U$ can be extended to at most one first order valuation $v$ of $E^U$, also $v_0$ can always be extended to at least one first order valuation $v$ of $E^U$ by the following considerations (which are obvious analogues of the propositional case). By a *valuation* tree for $A$ (with respect to the universe U) is meant a formation tree for $A$ together with an assignment of truth values to all points of $\mathscr{T}$ such that the truth values of each point (other than an end point) is determined from the truth values of its successors by the conditions $F_1, F_2$. (Cf. definition of a first order valuation). One easily shows by induction on the degree of $A$ that for any atomic valuation $v_0$ there exists exactly one valuation tree for $A$ whose end points receive the values given by $v_0$; in this valuation tree, $A$ receives a unique value. We define $v(A)$ to be this value. Thus $v$ is a valuation of the entire set $E^U$, and it is easily seen to be a first-order valuation. N.B. Unlike the case for propositional logic, valuation trees do not provide any *effective* method for determining the truth value of a formula under a given atomic valuation, since (in general) some points of the tree have *infinitely* many successors, so we can not in general actually *know* the truth value of such a point unless we know the truth-values of its infinitely many successors. However, even though the valuation procedure is non-effective it is mathematically well defined. [To formalize the argument would require using a language stronger than first order logic; it can be done, e.g. in set theory, or in the language known as second-order arithmetic.]

*Interpretations.* In propositional logic, we used the term "interpretation" to mean an atomic valuation. In first order logic, the term "interpretation" is traditionally used in the following different (but very closely related) sense.

We let $E$ be the set of all pure closed formulas of quantification theory. By an *interpretation* $I$ of $E$ in a universe $U$ is meant a function which assigns to each $n$-ary predicate $P$ an $n$-place relation $P^*$ of elements of $U$. An *atomic* $U$-sentence $P\mathcal{E}_1, ..., \mathcal{E}_n$ is called *true* under $I$ if the $n$-tuple $\mathcal{E}_1, ..., \mathcal{E}_n$ stands in the relation $P^*$. In this manner, the interpretation $I$ *induces* a unique atomic valuation $v_0$. And we say that an arbitrary element of $E^U$ (not necessarily atomic) is true under the interpretation $I$ if it is true under the induced atomic valuation $v_0$.

Conversely, if we start with an atomic valuation $v_0$ on a universe $U$, we associate with it that interpretation which defines each $P^*$ as the set of all $n$-tuples $\mathcal{E}_1, ..., \mathcal{E}_n$ such that $P\mathcal{E}_1, ..., \mathcal{E}_n$ is true under $v_0$. It is obvious that this interpretation $I$ induces $v_0$ back again, and is the *only* interpretation which induces $v_0$.

Thus, there is no essential difference between the points of view of interpretations and of atomic valuations, and we shall use whichever notion is more convenient at the moment.

*Example.* Let $U$ be the domain of natural numbers. Consider the formula $(\forall x)(\exists y)Pxy$. It is meaningless to ask whether this formula is true or false before we are given an interpretation which tells us what $P^*$ is. Suppose we interpret $Px, y$ to mean "$x$ is less than $y$"—i.e., suppose we consider any interpretation in which $P^*$ is the relation $<$. Under such an interpretation the above formula is clearly true. On the other hand, if we interpret $P$ to be the greater than relation, then the formula is false.

We have so far been using the letter "$E$" to denote the set of formulas built from *all* predicates of quantification theory. We shall sometimes consider the case when $E$ is rather the set of all formulas built from some *subset* of the predicates, in which case the definitions of $E^U$, *atomic valuation* of $E^U$, *first order valuation* of $E^U$ and *interpretation* of $E$ in U are the same as before (except that we take into consideration only those predicates which occur in at least one element of $E^U$).

*Example.* Let $P$ be a predicate of degree 1 and $Q$ be a predicate of degree 2. Let $E$ be the set of all formulas containing no predicates other than $P$ and $Q$. Let U be the *finite* universe $\{1,2,3\}$. The following is an example of an atomic valuation of $E^U$:

| True | False |
|------|-------|
| $P1$ | $P2$ |
| $P3$ | $Q1,1$ |
| $Q1,2$ | $Q2,2$ |
| $Q1,3$ | $Q3,3$ |
| $Q2,3$ | $Q2,1$ |
| | $Q3,1$ |
| | $Q3,2$ |

The truth value of *any* element of $E^U$ is now completely determined by the above table (since we have specified which *atomic* elements are true and which are false). We have thus determined a *first order* valuation defined on the entire set $E^U$.

Alternatively, we could have determined $v$ by the following *interpretation* (of the predicates $P, Q$):

$P$: The set $\{1,3\}$.

$Q$: The set of ordered pairs
   $\{\langle 1,2 \rangle, \langle 1,3 \rangle, \langle 2,3 \rangle\}$

equivalently the *less than* relation restricted to U.

*Models.* An interpretation $I$ in which every element of a given set $S$ is true is sometimes call a *model* of the set $S$.

*Validity and Satisfiability.* A pure formula $A$ is called *valid* if it is true under every interpretation (in every universe). Equivalently, $A$ is valid if for every universe U, $A$ is true under every first order valuation of $E^U$.

A formula $A$ is called (first order) *satisfiable* if it is true in at least one interpretation in at least one universe—equivalently if for at least one universe U there is at least one first order valuation in which $A$ is true. More generally, a set $S$ of pure formulas is called *satisfiable* (more generally, simultaneously first order satisfiable) if there is at least one interpretation in which all elements of $S$ are true.

One sometimes speaks of validity and satisfiability within a specific universe U. A formula is called *valid in* U if it is true in all interpretations in U, and satisfiable in U if it is true in at least one interpretation in U. And a set $S$ is called satisfiable in U if there is at least one interpretation in U which satisfies every element of $S$.

Thus $A$ is valid iff $A$ is valid in every universe; $A$ is satisfiable iff it is satisfiable in at least one universe.

Löwenheim proved the celebrated result that if $A$ is satisfiable at all, then $A$ must be satisfiable in some *denumerable* domain. Skolem subsequently extended this result, and showed that if a denumerable set $S$ of closed formulas is simultaneously satisfiable, then it is simultaneously satisfiable in a denumerable domain. This result (the so-called "Skolem-Löwenheim theorem") has a profound impact on the foundations of mathematics. We shall subsequently consider several proofs of this theorem.

*Exercise 1.* Show that the validity or satisfiability of a formula in a universe U depends only on the cardinality of U.

*Exercise 2.* Show that if a formula is satisfiable in U, then it is satisfiable in any larger domain (i. e. in any superset of U). Show that if $A$ is valid in U, then it is valid in any smaller domain (subset of U).

*Exercise 3.* Construct an example of a formula satisfiable in a denumerable universe, but not in any finite universe.

*Exercise 4.* Show that $A$ is valid iff $\sim A$ is not satisfiable. Show that $A$ is satisfiable iff $\sim A$ is not valid.

*Sentences with Parameters.* We have so far defined validity and satisfiability just for *pure* sentences (sentences with no parameters). Consider now a sentence $A(a_1, \ldots, a_n)$ containing exactly the parameters $a_1, \ldots, a_n$. For any universe U and any elements $k_1, \ldots, k_n$ of U by $A(k_1, \ldots, k_n)$ we mean the result of simultaneously substituting $k_1$ for $a_1, \ldots, k_n$ for $a_n$ in $A(a_1, \ldots, a_n)$. Given an interpretation $I$ of the predicates of $A(a_1, \ldots, a_n)$ in the universe U, we say that $A(a_1, \ldots, a_n)$ is *satisfiable under $I$* if there exists at least one $n$-tuple $\langle k_1, \ldots, k_n \rangle$ of U such that $A(k_1, \ldots, k_n)$ is true under $I$, and we say that $A(a_1, \ldots, a_n)$ is *valid under $I$* if for every $k_1, \ldots, k_n$ of U, $A(k_1, \ldots, k_n)$ is true under $I$. Then we say $A$ is valid (satisfiable) in U if $A$ is valid (satisfiable) under all interpretations (at least one interpretation) in U. And we say that $A(a_1, \ldots, a_n)$ is valid (satisfiable) if $A(a_1, \ldots, a_n)$ is valid (satisfiable) in all (at least one) universe.

The definition of simultaneous satisfiability of a set $S$ of sentences with parameters should now be obvious. Consider any mapping $\varphi$ from the set of parameters of $S$ (i.e. which occur in at least one element of $S$) into a universe U. For any $A \in S$, by $A^{\varphi}$ we mean the result of substituting for each parameter $a_i$ of $A$ its image $\varphi(a_i)$ under $\varphi$. Now we say that $S$ is (simultaneously) satisfiable in U if there exists an interpretation $I$ of the predicates of $S$ and there exists a "*substitution*" $\varphi$ mapping the parameters of $S$ into elements of U such that for any $A \in S$, $A^{\varphi}$ is true under $I$.

*Exercise.* Let $x_1, ..., x_n$ be variables which do not occur in $A(a_1, ..., a_n)$; let $A(x_1, ..., x_n)$ be the result of substituting $x_1$ for $a_1, ..., x_n$ for $a_n$. Show that $A(a_1, ..., a_n)$ is valid iff the pure sentence $(\forall x_1) \cdots (\forall x_n)$ $A(x_1, ..., x_n)$ is valid, and that $A(a_1, ..., a_n)$ is satisfiable iff the pure sentence $(\exists x_1), ..., (\exists x_n) A(x_1, ..., x_n)$ is satisfiable.

## § 3. Boolean Valuations vs. First-Order Valuations

By a *Boolean atom* we shall mean a sentence which is not the negation of any other sentence nor the conjunction, disjunction or conditional of other sentences. Equivalently, a sentence is a Boolean atom iff it is either an *atomic* sentence $P a_1, ..., a_n$ or is of one of the forms $(\forall x)A$ or $(\exists x)A$.

Consider now the universe $\vee$ whose elements are the parameters themselves. We can talk about *Boolean valuations* of $E^\vee$ and also about *first order valuations* of $E^\vee$; these are not the same thing. Obviously every first order valuation of $E^\vee$ is also a Boolean valuation of $E^\vee$, but a Boolean valuation of $E^\vee$ may fail to be a first order valuation of $E^\vee$.

All our results about Boolean valuations of propositional logic go over almost intact to *Boolean valuations* in quantification theory. For example, any assignment of truth values to all *Boolean atoms* of $E^\vee$ can be extended to one and only one Boolean valuation of $E^\vee$ (but in general this Boolean valuation fails to be a first-order valuation). Also the following version of the compactness theorem can be proved exactly as in propositional logic: If $S$ is an infinite subset of $E^\vee$ such that every finite subset of $S$ is *truth-functionally satisfiable* (i.e. true in at least one Boolean valuation of $E^\vee$) then the whole set $S$ is *truth-functionally satisfiable*. [This can be proved by any of the methods of Chapter III, everywhere replacing "propositional variable" by "Boolean atom"]. There is, however, another compactness theorem for first order logic— a much deeper result which we shall subsequently prove—which says that if every finite subset of $S$ is *first-order* satisfiable (i.e. true in at least one *first order* valuation) then $S$ is *first-order* satisfiable.

We must similarly distinguish *valid sentences* from *tautologies*. A valid sentence is true under all *first-order* valuations; a tautology is true under all *Boolean valuations* (even those which are not first order valuations). For example, the following sentence is not only valid, but even a tautology. $[(\forall x)Px \wedge (\forall x)Q x] \supset (\forall x)Px$. This sentence is of the form $(p \wedge q) \supset p$, which is a tautology of propositional logic.

However, the following sentence, though valid is *not* a tautology:

$$(\forall x)[P x \wedge Q x] \supset (\forall x)Px.$$

*Exercise 1.* Why is the last sentence not a tautology?

*Exercise 2.* Give an example of a sentence which is truth-functionally satisfiable but not first order satisfiable.

*Exercise 3.* [*Important!*] Show that a *quantifier-free* sentence (i.e. a sentence with no quantifiers) which is truth-functionally satisfiable must also be first order satisfiable.

Show therefore that if a *quantifier-free* sentence is valid, then it must be a tautology [this is a semantic version of Hilbert's first $\epsilon$-theorem]. More generally, show that any $S$ (finite or infinite) of *quantifier-free* sentences which is (simultaneously) truth-functionally satisfiable is also (simultaneously) first order satisfiable [Hint: Take any Boolean valuation of $E$ which satisfies $S$. This Boolean valuation induces a certain interpretation $I$ of the predicates of $S$ (in a manner we have discussed earlier). Show that every element $X$ of $S$ is true under this interpretation $I$ (use induction on the degree of $X$)].

## Chapter V

# First-Order Analytic Tableaux

### § 1. Extension of Our Unified Notation

We use "$\alpha$", "$\beta$" exactly as we did for propositional logic (only now construing "formula" to mean "closed formula of quantification theory"). We now add two more categories $\gamma$ and $\delta$ as follows.

For the moment let us work with unsigned formulas. Then "$\gamma$" shall denote any formula of one of the two forms $(\forall x)A$, $\sim(\exists x)A$, and for any parameter $a$, by $\gamma(a)$ we mean $A_a^x$, $\sim A_a^x$ respectively.

We use "$\delta$" to denote any formula of one of the two forms $(\exists x)A$, $\sim(\forall x)A$, and by $\delta(a)$ we respectively mean $A_a^x$, $\sim(A_a^x)$. We refer to $\gamma$-formulas as of *universal* type, and $\delta$-formulas as of *existential* type.

In working with signed formulas, $\gamma$ shall be any signed formula of one of the forms $T(\forall x)A$, $F(\exists x)A$, and $\gamma(a)$ is respectively $TA_a^x$, $FA_a^x$. And $\delta$ shall be any signed formula of one of the forms $T(\exists x)A$, $F(\forall x)A$, and $\delta(a)$ is respectively $TA_a^x$, $FA_a^x$.

In considering sentences with constants in the universe $U$, we use $\gamma$, $\delta$ in the same manner and for any $k \in U$, we define $\gamma(k)$, $\delta(k)$ similarly.

Under any interpretation in a universe $U$, the following facts clearly hold:

$F_1$: $\alpha$ is true iff $\alpha_1, \alpha_2$ are both true.
$F_2$: $\beta$ is true iff at least one of $\beta_1, \beta_2$ is true.
$F_3$: $\gamma$ is true iff $\gamma(k)$ is true for *every* $k \in U$.
$F_4$: $\delta$ is true iff $\delta(k)$ is true for *at least one* $k \in U$.

As consequences of the above facts, we have the following laws concerning satisfiability, of which $G_1, G_2, G_3$ are immediate and $G_4$ (which the reader should look at most carefully) we will prove. In these laws, $S$ is any set of formulas perhaps with parameters (but no other constants), and likewise with $\alpha$, $\beta$, $\gamma$, $\delta$. And "satisfiable" means *first order satisfiable*.

$G_1$: If $S$ is satisfiable, and $\alpha \in S$, then $\{S, \alpha_1, \alpha_2\}$ is satisfiable.

$G_2$: If $S$ is satisfiable and $\beta \in S$, then at least one of the two sets $\{S, \beta_1\}$, $\{S, \beta_2\}$ is satisfiable.

$G_3$: If $S$ is satisfiable and $\gamma \in S$, then, for *every* parameter $a$, the set $\{S, \gamma(a)\}$ is satisfiable.

$G_4$: If $S$ is satisfiable and $\delta \in S$, *and if a is any parameter which occurs in no element of S*, then $\{S, \delta(a)\}$ is satisfiable.

We leave the verification of $G_1, G_2, G_3$ to the reader; we shall now prove the very critical law $G_4$.

By hypothesis, there is an interpretation $I$ of all predicates of $S$ in some universe $U$ and a mapping $\varphi$ of all parameters of $S$ into elements of $U$ such that for every $A \in S$, the $U$-sentence $A^\varphi$ is true under $I$. In particular, $\delta^\varphi$ is true under $I$. The sentence $\delta^\varphi$ is a sentence with no parameters but with constants in $U$ and it is a sentence of existential type, call it $\delta_1$. Since $\delta_1$ is true under $I$, then (by $F_4$), there must be at least one element $k$ of $U$ such that $\delta_1(k)$ is true under $I$. Now $\varphi$ is defined on all parameters of $\{S, \delta(a)\}$, except for the parameter $a$. We extend $\varphi$ by defining $\varphi(a) = k$—call this extension $\varphi^*$. Then $\varphi^*$ is defined on *all* parameters of $\{S, \delta(a)\}$. Clearly, for every $A \in S, A^{\varphi^*}$ is the same expression as $A^\varphi$, so $A^{\varphi^*}$ is true under $I$. And $[\delta(a)]^{\varphi^*}$ is the same sentence as $\delta_1(k)$, hence $[\delta(a)]^{\varphi^*}$ is true under $I$. Hence, for every $A \in \{S, \delta(a)\}, A^{\varphi^*}$ is true under $I$. Thus $\{S, \delta(a)\}$ is satisfiable.

## § 2. Analytic Tableaux for Quantification Theory

Whether we work with signed formulas or not, our tableaux rules for first order logic are the following four:

*Rule A:* $\dfrac{\alpha}{\substack{\alpha_1 \\ \alpha_2}}$          *Rule B:* $\dfrac{\beta}{\beta_1 \mid \beta_2}$

*Rule C:* $\dfrac{\gamma}{\gamma(a)}$, where $a$ is *any* parameter.

*Rule D:* $\dfrac{\delta}{\delta(a)}$, where $a$ is a *new* parameter.

Rules $A$, $B$ are the same as in propositional logic. The new rules $C$, $D$ (for eliminating quantifiers), are *direct* rules; the only one of the four rules which is a branching rule is Rule $B$. In *signed* notation, Rules $C$, $D$ are as follows:

*Rule C:* $\quad \dfrac{T(\forall x)A}{T A_a^x} \quad \dfrac{F(\exists x)A}{F A_a^x}$

*Rule D:* $\quad \dfrac{T(\exists x)A}{T A_a^x}$ , with proviso (that $a$ is new)

$\qquad\qquad \dfrac{F(\forall x)A}{F A_a^x}$ , with same proviso

Using *unsigned* formulas, our quantificational rules are:

*Rule C:* $\quad \dfrac{(\forall x)A}{A_a^x} \quad \dfrac{\sim(\exists x)A}{\sim A_a^x}$

*Rule D:* $\quad \dfrac{(\exists x)A}{A_a^x}$ , with proviso

$\qquad\qquad \dfrac{\sim(\forall x)A}{\sim A_a^x}$ , with proviso

*Discussion concerning Rule D.* This rule is a formalization of the following informal argument used constantly in mathematics. Suppose in the course of an argument we have proved that there exists an element $x$ having a certain property $P$—i.e. we have proved the statement $(\exists x)Px$. We then say, "let $a$ be such an $x$" and we write $Pa$. Of course, we are not asserting that $P$ holds for *every* $a$, but just for at least one. If we subsequently show that for another property $Q$, there exists an $x$ such that $Qx$, we cannot legitimately say "let $a$ be such an $x$", because we have already committed the symbol "$a$" to being the name of some $x$ such that $Px$ and we do not know that there is any single $x$ having both the properties $P$ and $Q$. Thus we take a *new* parameter $b$, and say "let $b$ be such an $x$", and we write $Qb$. This is the reason for the proviso in Rule $D$.

Actually we can liberalize Rule $D$ by replacing the clause "providing $a$ is new", by "providing $a$ is new, or else $a$ has not been previously introduced by Rule $D$, and does not occur in $\delta$, and no parameter of $\delta$ has been previously introduced by Rule $D$". Under this liberalization, proofs can sometimes be shortened (cf. Example 2 below.)

The idea behind this liberalization is this. Suppose in the course of an argument we prove a sentence $(\forall x)Px$ (which is of type $\gamma$). Then we conclude $Pa$. We have not really committed "$a$" to being the name of any particular individual; $Pa$ holds for *every* value of $a$. So if we subsequently prove a sentence $(\exists x)Qx$, we can legitimately say, "let $a$ be such an $x$", and for the same value of $a$, $Pa$ will also hold.

The above discussion is but an informal foreshadowing of a precise argument showing the *consistency* of the tableau method for first order logic. Actually, if we stick to the *strict* version of Rule $D$, the consistency is almost immediate from the conditions $G_1$, $G_2$, $G_3$, $G_4$ of satisfiability which we stated in § 1.

For suppose $\theta$ is a branch of a tableau and that $\theta$ is satisfiable. If we extend $\theta$ by rule $A$, $C$ or $D$ then the resulting extension is again satisfiable (by $G_1$, $G_3$, $G_4$, respectively). If we simultaneously extend $\theta$ to 2 branches $\theta_1$, $\theta_2$ by one application of rule $B$, then at least one of $\theta_1$, $\theta_2$ is again satisfiable (by $G_2$). Thus any immediate extension of a tableau which is satisfiable (in the sense that at least one of its branches is satisfiable) is again satisfiable. Therefore (by induction) if the origin of a tableau is satisfiable, then at least one branch of the tableau is satisfiable and hence open. Therefore if a tableau closes, then the origin is indeed unsatisfiable—stated otherwise, every provable sentence is valid.

The precise justification of the *liberalized* version of Rule $D$ is a bit more delicate; we shall return to this later. Meanwhile we wish to get on with some concrete working familarity with First Order Tableaux.

*Example 1.* The following tableau is a proof of the sentence

$$(\forall x)[Px \supset Qx] \supset [(\forall x)Px \supset (\forall x)Qx]$$

| | | | | |
|---|---|---|---|---|
| (1) | $\sim[(\forall x)[Px \supset Qx] \supset [(\forall x)Px \supset (\forall x)Qx]]$ | | | |
| (2) | $(\forall x)(Px \supset Qx)$ | | | (1) |
| (3) | $\sim[(\forall x)Px \supset (\forall x)Qx]$ | | | (1) |
| (4) | $(\forall x)Px$ | | | (3) |
| (5) | $\sim(\forall x)Qx$ | | | (3) |
| (6) | $\sim Qa$ | | | (5) |
| (7) | $Pa$ | | | (4) |
| (8) | $Pa \supset Qa$ | | | (2) |
| (9) | $\sim Pa$ | (8) | (10) | $Qa$ | (8) |
| | $\times$ | | | $\times$ |

*Example 2.* We wish to give 2 different proofs of the sentence $(\exists y)[(\exists x)Px \supset Py]$. The first proof uses the *strict* form of Rule $D$, and the second (which is shorter) uses the *liberalized* version of Rule $D$:

**Proof 1.**          (1)   $F(\exists y)[(\exists x)Px \supset Py]$

                      (2)   $F(\exists x)Px \supset Pa$                (1)

                      (3)   $T(\exists x)Px$                          (2)

                      (4)   $FPa$                                     (2)

                      (5)   $TPb$                                     (3)

                      (6)   $F(\exists x)Px \supset Pb$               (1)

                      (7)   $FPb$

                             x

**Proof 2.**          (1)   $F(\exists y)[(\exists x)Px \supset Py]$

                      (2)   $F(\exists x)Px \supset Pa$                (1)

                      (3)   $T(\exists x)Px$                          (2)

                      (4)   $FPa$                                     (2)

                      (5)   $TPa$                                     (3)

                             x

*Exercises.*  Prove the following formulas:

$$(\forall y)[(\forall x)Px \supset Py]$$

$$(\forall x)Px \supset (\exists x)Px$$

$$(\exists y)[Py \supset (\forall x)Px]$$

$$\sim(\exists y)Py \supset [(\forall y)((\exists x)Px \supset Py)]$$

$$(\exists x)Px \supset (\exists y)Py$$

$$(\forall x)[Px \wedge Qx] \equiv (\forall x)Px \wedge (\forall x)Qx$$

$$[(\forall x)Px \vee (\forall x)Qx] \supset (\forall x)[Px \vee Qx]$$

(the converse is not valid!)

$$(\exists x)(Px \vee Qx) \equiv ((\exists x)Px \vee (\exists x)Qx)$$

$$(\exists x)(Px \wedge Qx) \supset ((\exists x)Px \wedge (\exists x)Qx)$$

(the converse is not valid).

In the next group, $C$ is any *closed* formula—or at least the variable $x$ does not occur free in it:

$$(\forall x)[Px \vee C] \equiv [(\forall x)Px \vee C]$$

$$(\exists x)[Px \wedge C] \equiv [(\exists x)Px \wedge C]$$

$$(\exists x)C \equiv C$$

$$(\forall x)C \equiv C$$

$$(\exists x)[C \supset Px] \equiv [C \supset (\exists x)Px]$$

$$(\exists x)[Px \supset C] \equiv [(\forall x)Px \supset C]$$

$$(\forall x)[C \supset Px] \equiv [C \supset (\forall x)Px]$$

$$(\forall x)[Px \supset C] \equiv [(\exists x)Px \supset C]$$

*Show* $(H \wedge K) \supset L$, where
$$H = (\forall x)(\forall y)[Rxy \supset Ryx] \quad (R \text{ is } symmetric)$$
$$K = (\forall x)(\forall y)(\forall z)[(Rxy \wedge Ryz) \supset Rxz] \quad (R \text{ is } transitive)$$
$$L = (\forall x)(\forall y)[Rxy \supset Rxx] \quad (R \text{ is } reflexive \text{ on its domain of}$$
$$\text{definition)}.$$

For a *hard* one, try the following exercise (taken from Quine [1]):
Show $(A \wedge B) \supset C$, where
$$A = (\forall x)[(Fx \wedge Gx) \supset Hx] \supset (\exists x)[Fx \wedge \sim Gx]$$
$$B = (\forall x)[Fx \supset Gx] \vee (\forall x)[Fx \supset Hx]$$
$$C = (\forall x)[(Fx \wedge Hx) \supset Gx] \supset (\exists x)[Fx \wedge Gx \wedge \sim Hx]$$

## § 3. The Completeness Theorem

Now we turn to the proof of one of the major results in quantification theory: Every valid sentence is provable by the tableau method.

This is a form of Gödel's famous completeness theorem. Actually Gödel proved the completeness of a different formalization of Quantification Theory, but we shall later show how the completeness of the tableau method implies the completeness of the more conventional formalizations. The completeness proof we now give is along the lines of Beth [1] or Hintikka [1]—and also Anderson and Belnap [1].

Let us first briefly review our completeness proof for tableaux in *propositional* logic, and then see what modifications will suggest themselves. In the case of propositional logic, we reach a completed tableau after finitely many stages. Upon completion, every open branch is a Hintikka set. And by Hintikka's lemma, every Hintikka set is truth-functionally satisfiable.

Our first task is to give an appropriate definition of "Hintikka set" for first order logic in which we specify conditions not only on the $\alpha$'s and $\beta$'s but also on the $\gamma$'s and $\delta$'s as well. We shall define Hintikka sets for arbitrary universes $U$ of constants.

**Definition.** By a Hintikka set (for a universe $U$) we mean a set $S$ (of $U$-formulas) such that the following conditions hold for every $\alpha$, $\beta$, $\gamma$, $\delta$ in $E^U$:

$H_0$: No *atomic* element of $E^U$ and its negation (or conjugate, if we are working with signed formulas) are both in $S$.

$H_1$: If $\alpha \in S$, then $\alpha_1$, $\alpha_2$ are both in $S$.

$H_2$: If $\beta \in S$, then $\beta_1 \in S$ or $\beta_2 \in S$.

$H_3$: If $\gamma \in S$, then for *every* $k \in U$, $\gamma(k) \in S$.

$H_4$: If $\delta \in S$, then for at least one element $k \in U$, $\delta(k) \in S$.

Now we show

**Lemma** *(Hintikka's lemma for first order logic). Every Hintikka set S for a domain U is (first order) satisfiable—indeed in the domain U.*

**Proof.** We must find an *atomic* valuation of $E^U$ in which all elements of $S$ are true. We do this exactly as we did propositional logic—viz., for every *atomic* sentence $P\xi_1, ..., \xi_n$ of $E^U$, give it the value $t$ if $TP\xi_1, ..., \xi_n$ is an element of $S$, $f$ if $FP\xi_1, ..., \xi_n$ is an element of $S$, and either $t$ or $f$ at will if neither is an element of $S$. We must show that each element $X$ of $S$ is true under this atomic valuation. Again we do this by induction on the degree of $X$. If $X$ is of degree 0, it is immediate that $X$ is true (under this valuation). Now suppose that $X$ is of positive degree and that every element of $S$ of lower degree is true. We must show $X$ is true. Since $X$ is not of degree 0, then it is either some $\alpha$, $\beta$, $\gamma$ or $\delta$. If it is an $\alpha$ or a $\beta$, then it is true for exactly the same reasons as in the proof of Hintikka's lemma for propositional logic (viz. if it is an $\alpha$, then $\alpha_1$, $\alpha_2$ are both in $S$, hence both true (by induction hypothesis), hence $\alpha$ is true; if it is a $\beta$, then at least one of $\beta_1$, $\beta_2$ is in $S$, and hence true, so $\beta$ is true). Thus the new cases to consider are $\gamma$, $\delta$.

Suppose it is a $\gamma$. Then, for *every* $k \in U$, $\gamma(k) \in S$ (by $H_3$), but every $\gamma(k)$ is of lower degree than $\gamma$, hence true by inductive hypothesis. Hence $\gamma$ must be true.

Suppose it is a $\delta$. Then for at least one $k \in U$, $\delta(k) \in S$ (by $H_4$). Then $\delta(k)$ is true by inductive hypothesis, hence $\delta$ is true.

We next consider how we can use Hintikka's lemma for our completeness proof. In propositional logic, tableaux terminate after finitely many steps. But a tableau for first order logic may run on infinitely without ever closing. Suppose this should happen. Then we generate an infinite tree $\mathscr{T}$, and by König's Lemma, $\mathscr{T}$ contains an infinite branch $\theta$. Clearly $\theta$ is open, but do the elements of $\theta$ necessarily constitute a Hintikka set? The answer is "no" as the following considerations will show.

For any $X$ on a branch $\theta$ of degree $> 0$, define $X$ to be *fulfilled* on $\theta$ if either: (i) $X$ is an $\alpha$, and $\alpha_1$, $\alpha_2$ are both on $\theta$; (ii) $X$ is a $\beta$ and at least one of $\beta_1$, $\beta_2$ is on $\theta$; (iii) $X$ is a $\gamma$ and for *every* parameter $a$, $\gamma(a)$ is on $\theta$; (iv) $X$ is a $\delta$ and for at least one parameter $a$, $\delta(a)$ is on $\theta$. Now suppose $\mathscr{T}$ is a finite tableau and that a branch contains two $\gamma$-sentences—call them $\gamma_1$ and $\gamma_2$. Now suppose we use $\gamma_1$ and successively adjoin $\gamma_1(a_1)$, $\gamma_1(a_2), ..., \gamma_1(a_n), ...$, for *all* the parameters $a_1, a_2, ..., a_n, ....$ We thus generate an infinite branch and we have clearly taken care of fulfilling $\gamma_1$, but we have totally neglected $\gamma_2$. Or it is possible to fulfill a $\gamma$-formula on a branch but neglect one or several $\alpha$, $\beta$, or $\delta$ formulas on the branch. Thus there are many ways in which an infinite tableau can be generated without all—or even any—open branches being Hintikka sets. The key problem is to find a *systematic* procedure which will guarantee that any tableau constructed according to the procedure is such that if it runs on infinitely, every open branch will have to be a Hintikka sequence.

Many such procedures exist in the literature; the reader should at this point try to work out such a procedure for himself before reading further.

The following systematic procedure seems to be as simple and direct as any. In this procedure of generating the tree, at each stage certain points of the tree are declared to have been "used" (as a practical book-keeping device, we can put a check mark to the right of a point of the tableau as soon as we have used it).

Now for a precise description of the procedure. We start the tableau by placing the formula whose satisfiability we are testing at the origin. This concludes the first stage. Now suppose we have concluded the $n$th stage. Then our next act is determined as follows. If the tableau already at hand is closed, then we stop. Also, if every non-atomic point on every open branch of the tableau at hand has been used, then we stop. If neither, then we pick a point $X$ of *minimal* level (i.e. as high up on the tree as possible) which has not yet been used and which appears on at least one open branch[1]), and we extend the tableau at hand as follows: we take *every* open branch $\theta$ passing through the point $X$, and

1) If $X$ is an $\alpha$, we extend $\theta$ to the branch $(\theta, \alpha_1, \alpha_2)$.

2) If $X$ is a $\beta$ then we simultaneously extend $\theta$ to the 2 branches $(\theta, \beta_1)$ and $(\theta, \beta_2)$.

3) If $X$ is a $\delta$ then we take the first parameter $a$ which does not appear on the tree and we extend $\theta$ to $(\theta, \delta(a))$.

4) If $X$ is a $\gamma$ (and this is the delicate case!), then we take the first parameter $a$ such that $\gamma(a)$ does not occur on $\theta$, and we extend $\theta$ to $(\theta, \gamma(a), \gamma)$. (In other words we add $\gamma(a)$ as an endpoint to $\theta$ and then we *repeat* an occurrence of $\gamma$!)

Having performed acts 1—4 (depending on whether $X$ is respectively an $\alpha$, $\beta$, $\gamma$, $\delta$), we then declare the point $X$ to be used, and this concludes the stage $n+1$ of our procedure.

**Discussion.** To describe the above procedure more informally, we *systematically* work our way down the tree, automatically fulfilling all $\alpha$, $\beta$ and $\delta$ formulas which come our way. As to the $\gamma$-formulas, when we use an occurrence of $\gamma$ on a branch $\theta$ to subjoin an instance $\gamma(a)$, the purpose of *repeating* an occurrence of $\gamma$ is that we must sooner or later come down the branch $\theta$ and use this repeated occurrence, from which we adjoin another instance $\gamma(b)$ and repeat an occurrence of $\gamma$

---

[1]) If the reader wishes to make the procedure completely deterministic he can, e. g. pick the *leftmost* such unused point of minimal level.

again, which we in turn use again, and so forth. In this way, we are certain of fulfilling all $\gamma$ formulas (as well as the $\alpha$, $\beta$, and $\delta$ formulas).

There are two cases in which a first-order tableau can have a *finite* open branch on which every non-atomic point has been used (so that the branch is a Hintikka sequence). One is when no $\gamma$-formulas occur on the branch; the other is when each $\gamma$ that occurs has a vacuous quantifier, so that $\gamma(a)$ is the same sentence for every parameter $a$ and $\gamma$ has to be used only once.

We should also remark that strictly speaking, the tree generated by our procedure is not literally a tableau, since there are no tableau rules allowing for arbitrary repetitions of $\gamma$-formulas (or any other formulas for that matter). However it is trivial to verify that if we should subsequently delete these repetitions, then the resulting tree is indeed a tableau. Or to put it another way, if it is possible to close a tableau allowing arbitrary repetitions, then we could close the tableau without these arbitrary repetitions (for anything subjoined using these repetitions could just as well have come from the originals higher up on the tree).

We shall use the term "systematic tableau", to mean a tableau constructed by the above systematic procedure. By a *finished* systematic tableau we shall mean a systematic tableau which is either infinite, or else finite but cannot be extended further by continuing the systematic procedure (in other words, for each open branch, all non-atomic elements have already been used). From our discussion above, we have:

**Theorem 1.** *For any finished systematic tableau, every open branch is a Hintikka sequence ( for the denumerable universe V of parameters).*

From Theorem 1 and Hintikka's lemma for first order logic, we at once have:

**Theorem 2.** *In any finished systematic tableau $\mathcal{T}$, every open branch is simultaneously ( first order) satisfiable.*

Theorem 2 in turn yields:

**Theorem 3.** *(Completeness Theorem for First Order Tableaux). If X is valid, then X is provable—i.e., there exists a closed tableau for $FX$. Indeed, if X is valid, then the systematic tableau for $FX$ must close after finitely many steps.*

**Proof.** Suppose $X$ is valid. Let $\mathcal{T}$ be the finished systematic tableau starting with $FX$. If $\mathcal{T}$ contained an open branch $\theta$ then by theorem 2, $\theta$ would be satisfiable, hence $FX$, being a term of $\theta$, would be satisfiable, contrary to the hypothesis that $X$ is valid. Thus $X$ is provable.

Concerning the second statement, by König's lemma, a closed infinite tableau is impossible, because if $\mathcal{T}$ is closed, then every branch of $\mathcal{T}$ is of finite length, hence $\mathcal{T}$ must be finite.

Theorem 2 also yields:

**Theorem 4.** *(Löwenheim's Theorem). If X is satisfiable at all, then X is satisfiable in a denumerable domain.*

**Proof.** Let $\mathcal{T}$ be a finished systematic tableau starting with $X$. We have shown earlier that if $\mathcal{T}$ were closed, then $X$ could not be satisfiable. Hence $\mathcal{T}$ contains at least one open branch $\theta$. Then by theorem 2, there is an interpretation in a *denumerable* domain in which all elements of $\theta$—in particular $X$ itself—are true. This concludes our proof.

*Discussion of Systematic Tableaux.* In general, systematic tableaux may take much longer to close than a tableau constructed using some ingenuity. The whole point of considering systematic tableaux is that they are *bound* to close if the origin is unsatisfiable, whereas a tableau constructed at random may fail to close even though the origin is unsatisfiable. Yet it is still true that if $X$ is unsatisfiable, a cleverly constructed tableau for $X$ may close faster than the systematic tableau. Thus if the reader is in a very lazy mood, and if he wishes to test the satisfiability of $X$, he can mechanically run a *systematic* tableau for $X$, confident in the knowledge that if $X$ is unsatisfiable, then the tableau will eventually close. On the other hand, if he is in a brighter and more creative mood, then he can be alert and seize the first opportunity—or indeed plan clever strategies—for closing the tableau quickly, even though he disregards the systematic procedure.

We assume the reader has already worked several of the exercises at the end of § 2. It is highly unlikely that any of the tableaux constructed therein are systematic. At this point, the reader should really try one or two of these exercises using *systematic* tableaux, and should compare the length of the proof with the non-systematic tableau he has already constructed.

The particular systematic procedure we have given is by no means the best as a practical proof procedure. The following procedure, though a bit more difficult to justify, is better from the viewpoint of getting shorter proofs.

Instead of always using the *highest* unused point of a given branch, do the following: At any given stage first use all the $\alpha$ and $\delta$ points. (This clearly must terminate after finitely many steps.) Then use up all the $\beta$ points (and again this must terminate after finitely many steps). *Then* use a $\gamma$-point of maximal height (i.e., maximal level) in the manner indicated in the first procedure.

A few working examples will convince the reader of the practical superiority of this procedure. Needless to say, it is still capable of many further improvements; such a study is a subject in itself known as "mechanical theorem proving". This subject—which is of relatively recent origin—investigates proof procedures from the viewpoint of speed. These procedures are sometimes programmed on computing machines.

*Atomic Closure.* As in propositional logic, we call a tableau *atomically closed* if every branch contains some *atomic* element and its conjugate. If we modify our systematic procedure by replacing "closed" by "atomically closed" and "open" by "atomically open" (meaning "not atomically closed"), then for any finished tableau constructed by this modified procedure, each branch is either *atomically* closed or is a Hintikka set. Thus if $X$ is unsatisfiable then there not only exists a closed tableau for $X$, but even an atomically closed tableau for $X$. This fact will be important later on.

Again we proved the above fact using a model-theoretic argument. But it is equally simple to give a purely systematic proof that any closed tableau can be further extended to an atomically closed tableau; viz. by showing by induction on $X$ that there is an atomically closed tableau for any finite set which contains both $X$ and $\bar{X}$. Thus any closed branch of a tableau can be extended to an atomically closed tableau. Therefore, any closed tableau can be extended to an atomically closed tableau. [In verifying this last statement, do not overlook what happens with Rule $D$!]

*Satisfiability in a Finite Domain.* Suppose that in constructing a tableau, we reach a stage in which the tableau is not closed, yet there is at least one branch $\theta$ whose elements constitute a Hintikka set *for the finite domain of those parameters which occur in at least one element of $\theta$*. It is then pointless to continue the tableau further, for by Hintikka's Lemma, the set $\theta$ (and hence in particular the origin) is satisfiable in a *finite* domain.

*Example.* We have asserted that the formula $[(\forall x)Px \vee (\forall x)Qx] \supset (\forall x)[Px \vee Qx]$ is valid, but that its converse $(\forall x)[Px \vee Qx] \supset [(\forall x)Px \vee (\forall x)Qx]$ is not valid. In other words the signed formula $F(\forall x)[Px \vee Qx] \supset [(\forall x)Px \vee (\forall x)Qx]$ is satisfiable. Let us show this by the tableau method. Consider the following tableau:

$$
\begin{array}{lll}
(1) & F(\forall x)(Px \vee Qx) \supset ((\forall x)Px \vee (\forall x)Qx) & \\
(2) & T(\forall x)(Px \vee Qx) & (1) \\
(3) & F(\forall x)Px \vee (\forall x)Qx & (1) \\
(4) & \quad\quad F(\forall x)Px & (3) \\
(5) & \quad\quad F(\forall x)Qx & (3) \\
(6) & \quad\quad FPa & (4) \\
(7) & \quad\quad FQb & (5) \\
\end{array}
$$

| (8) | | | $TPa \lor Qa$ | | (2) | | | |
|---|---|---|---|---|---|---|---|---|
| (9) | | | $TPb \lor Qb$ | | (2) | | | |
| (10) | $TPa$ | (8) | (11) | $TQa$ | | | (8) | |
| | $x$ | | (12) | $TPb$ | (9) | (13) | $TQb$ | (9) |
| | | | | | | | $x$ | |

The above tableau has one open branch $\theta$. In this branch the $\alpha$-points (1), (3) are both fulfilled; the $\beta$-points (8), (9) are both fulfilled; the $\delta$-points (4), (5) are both fulfilled; and the $\gamma$-point (2) is fulfilled *for the 2 element domain* $\{a,b\}$. Then the proof of Hintikka's lemma shows us that all elements of $\theta$ are true under the following atomic valuation:

$$Pa \text{ — false}$$
$$Pb \text{ — true}$$
$$Qa \text{ — true}$$
$$Qb \text{ — false}$$

Equivalently, (1) is satisfied by the interpretation in which $P$ is the set whose only element is $b$, and $Q$ is the set whose only element is $a$. The reader can easily verify that all elements of $\theta$—in particular (1) itself—are true under this finite interpretation.

We thus see how tableaux not only can be used to show certain formulas to be unsatisfiable (or equivalently to show certain formulas to be valid), but also can sometimes be used to show certain formulas to be satisfiable (when these formulas happen to be satisfiable in a finite domain). The real "mystery class" consists of those formulas which are neither unsatisfiable nor satisfiable in any finite domain. If we construct a tableau—even a systematic one—for any such formula, the tableau will run on infinitely, and at no finite stage will we ever know that the formula is or is not satisfiable. There are formulas which are satisfiable but not in any finite domain (cf. exercise below). However, the demonstration of their satisfiability cannot be accomplished within the framework of analytic tableaux.

*Exercise:* Let $H$ be the conjunction of the following three sentences:

1) $(\forall x)(\exists y)(Rx,y)$
2) $\sim(\exists x)Rx,x$
3) $(\forall x)(\forall y)(\forall z)[(Rx,y \land Ry,z) \supset Rx,z]$

Clearly $H$ is true in the domain of the natural numbers if we interpret "$R$" to be the *less than* relation. Prove that $H$ is not satisfiable in any *finite* domain. [This solves Ex. 3, p. 50.]

## § 4. The Skolem-Löwenheim and Compactness Theorems for First-Order Logic

We have already proved Löwenheim's theorem that every satisfiable formula is satisfiable in a denumerable domain. Our proof rested on the method of *systematic* tableaux for a single formula. To prove the *Skolem-Löwenheim* theorem (viz. that every simultaneously satisfiable denumerable set $S$ is satisfiable in a denumerable domain), we must extend our method and consider systematic tableaux for *sets $S$* of formulas. We shall restrict ourselves for a while to *pure* sets $S$,—i.e., sets of closed formulas with no parameters. First of all, by a (first-order) tableau *for $S$* we mean a tree constructed as follows. We start by placing any element of $S$ at the origin. Then at any stage, we can either use Rule $A$, $B$, $C$, $D$ or we may adjoin any element of $S$ to the end of any open branch; the elements so adjoined shall sometimes be referred to as the *premises* of the tableau (we might think of the process of constructing a tableau for $S$ as a test for whether the assumption of all elements of $S$ as *premises* leads to a contradiction).

By a *complete* tableau for $S$ we mean a tableau for $S$ such that every open branch is a Hintikka set[1]) (for the universe $K$ of parameters) and also contains all elements of $S$. (In particular, every *closed* tableau for $S$ is vacuously a complete tableau for $S$.)

**Theorem 5.** *For any $S$, there exists a complete tableau for $S$.*

**Proof.** We modify our "systematic" construction of a tableau for a single formula as follows. We first arrange $S$ in some denumerable sequence $X_1, X_2, ..., X_n, ...$. Then we begin by placing $X_1$ at the origin. This concludes the first stage. Now suppose we have concluded the $n$th stage. Then we proceed exactly as before (i.e., as in the case of a systematic tableau for a single formula), but then before we conclude the $(n+1)$th stage, we adjoin $X_{n+1}$ to the end of every open branch.

This concludes our definition of a systematic tableau for $S$. It is obvious that such a tableau must be a complete tableau for $S$.

**Remark.** If $S$ should be a finite rather than a denumerable set, then, of course, we arrange $S$ in a finite sequence $X_1, ..., X_k$ and carry through the construction as above through the $k$th stage, and from then on, proceed as in the construction for a single formula.

We leave the proof of the following to the reader:

**Lemma.** *If there exists a closed tableau for (a pure) set $S$, then some finite subset of $S$—to wit, the set of premises of the tableau—is unsatisfiable[2]). Stated otherwise, if all subsets of $S$ are (first order) satisfiable, then no tableau for $S$ can close.*

---

[1]) When we refer to a *branch* $\theta$ (which is really a sequence rather than a set) as being a Hintikka set, we mean, of course, that the set of elements of $\theta$ constitutes a Hintikka set.

[2]) We remind the reader that if a tableau is closed, then (by König's lemma) it must be finite and hence its set of premises is finite.

We now have all the pieces necessary for the following theorem, which simultaneously yields the Skolem-Löwenheim theorem and the compactness theorem for first-order logic.

**Theorem 6.** *If all finite subsets of (a pure) set S are satisfiable then the entire set S is simultaneously satisfiable in a denumerable domain.*

**Proof.** By theorem 5, there exists a complete tableau $\mathcal{T}$ for $S$. By hypothesis and the above lemma, $\mathcal{T}$ cannot be closed; let $\theta$ be an open branch. Then $\theta$ is a Hintikka set (for the denumerable universe of parameters), and all elements of $S$ are terms of $\theta$. The result then follows by Hintikka's lemma.

Hintikka's lemma, together with Theorem 5, also yields:

**Theorem 7.** *If no tableau for S can close, then S is satisfiable in a denumerable domain.*

Chapter VI

# A Unifying Principle

At this point we wish to discuss a principle which we introduced in [2] and which simultaneously yields several of the major results in Quantification Theory. The *mathematical* content of this principle is not really very different from that of our last theorem (Theorem 7, Chapter V), but it is in a form which makes no reference to the particular formal system of tableaux (or to any other specific formal system, for that matter). We believe that this is a good point to discuss this principle while the tableau method is still fresh in the reader's mind. We shall apply the principle several times in the further course of this study.

## § 1. Analytic Consistency

The word "set" shall (in this chapter) mean set of *sentences* (of Quantification Theory). We shall let "$\Gamma$" denote any *property* of sets which is of *finite character* (which we recall means that a set $S$ has the property $\Gamma$ iff all finite subsets of $S$ have the property $\Gamma$). To avoid circumlocution (as well as for other reasons which will subsequently become manifest), we shall say that a set $S$ is $\Gamma$-*consistent* if $S$ has the property $\Gamma$, and $\Gamma$-*inconsistent* if $S$ does *not* have the property $\Gamma$.

Now we define $\Gamma$ to be an *analytic consistency property*[1]) (for first order logic) if for every $\Gamma$-consistent set $S$, the following conditions hold:

$A_0$: $S$ contains no *atomic* element and its conjugate (or no atomic element and its negation, if we are working with unsigned formulas).

$A_1$: If $\alpha \in S$, then $\{S, \alpha_1\}$ and $\{S, \alpha_2\}$ (and hence also $\{S, \alpha_1, \alpha_2\}$) are $\Gamma$-consistent.

$A_2$: If $\beta \in S$, then either $\{S, \beta_1\}$ is $\Gamma$-consistent or $\{S, \beta_2\}$ is $\Gamma$-consistent.

$A_3$: If $\gamma \in S$, then $\{S, \gamma(a)\}$ is $\Gamma$-consistent.

$A_4$: If $\delta \in S$, then $\{S, \delta(a)\}$ is $\Gamma$-consistent, provided $a$ does not occur in $S$.

It will sometimes be convenient to use conditions $A_0 - A_4$ in the following equivalent form:

$A'_0$: Any set containing an atomic element and its conjugate (negation) is $\Gamma$-inconsistent.

$A'_1$: If $\{S, \alpha, \alpha_1\}$ is $\Gamma$-inconsistent, or if $\{S, \alpha, \alpha_2\}$ is $\Gamma$-inconsistent, so is $\{S, \alpha\}$.

$A'_2$: If $\{S, \beta_1\}$ and $\{S, \beta_2\}$ are both $\Gamma$-inconsistent, so is $\{S, \beta\}$.

$A'_3$: If $\{S, \gamma, \gamma(a)\}$ is $\Gamma$-inconsistent, so is $\{S, \gamma\}$.

$A'_4$: If $\{S, \delta(a)\}$ is $\Gamma$-inconsistent, and if $a$ does not occur in $\{S, \delta\}$, then $\{S, \delta\}$ is $\Gamma$-inconsistent.

*Examples.* Suppose we define $\Gamma(S)$ to mean that all finite subsets of $S$ are satisfiable. It is easily verified that this property $\Gamma$ is an analytic consistency property. Another important example of an analytic consistency property $\Gamma$ is this: Define $S$ to be $\Gamma$-consistent if no analytic tableau for $S$ can close. This is quite trivially an analytic consistency property.

In what follows, we shall let "$\Gamma$" stand for any analytic consistency property. Suppose that a set $S$ is $\Gamma$-consistent, and that $\mathscr{T}$ is a finite tableau containing a branch $\theta$ which is $\Gamma$-consistent with $S$ (by which we mean that the union of $S$ with the set of terms of $\theta$ is $\Gamma$-consistent). If we extend $\theta$ to a branch $\theta^*$ by an application of Rule $A$, $C$ or $D$, then $\theta^*$ is again $\Gamma$-consistent with $S$ (by conditions $A_1, A_3, A_4$ respectively). If we extend $\theta$ to $\theta^*$ by adjoining an element of $S$ (as a premise of the tableau), then trivially, $\theta^*$ is again $\Gamma$-consistent with $S$. If we simultaneously extend $\theta$ to two branches $\theta_1, \theta_2$ by Rule $B$, then either $\theta_1$ is $\Gamma$-consistent with $S$ or $\theta_2$ is $\Gamma$-consistent with $S$ (by condition $A_2$). If we extend some other branch of $\mathscr{T}$ (by a tableau rule), then $\mathscr{T}$ still contains the branch $\theta$ which is $\Gamma$-consistent with $S$. It follows by an obvious induction

---

[1]) This differs in minor details from what we called an "abstract consistency property" in [2]. The present definition seems to us a slight improvement.

that if $S$ is $\Gamma$-consistent, then at any stage of the construction of a tableau for $S$, we must always have at least one branch $\theta$ which is $\Gamma$-consistent with $S$. Then this branch $\theta$ must certainly be open (by condition $A_0$). We have thus proved

**Lemma 1.** *For any pure set $S$, if $S$ is $\Gamma$-consistent (where $\Gamma$ is any analytic consistency property) then no tableau for $S$ can close.*

**Remark.** The lemma for Theorem 6 of the last chapter is a special case of the above lemma, taking for $\Gamma$ the property "all finite subsets are satisfiable".

The above lemma and Theorem 7 of the last chapter at once yield:

**Theorem I.** [*A Unifying Principle*]. *For any pure set $S$, if $S$ is $\Gamma$-consistent (where $\Gamma$ is any analytic consistency property), then $S$ is simultaneously satisfiable in a denumerable domain.*

*Exercise.* Prove Theorem I for a set $S$ not necessarily pure but such that only finitely many parameters occur in $S$.

Let us now call an *unsigned* sentence $X$ $\Gamma$-*provable* if the unit set $\{F X\}$ is $\Gamma$-inconsistent. Theorem I at once yields

**Corollary.** *If $X$ is valid, then $X$ is $\Gamma$-provable.*

## § 2. Further Discussion of Analytic Consistency

We have proved Theorem I using analytic tableaux, yet Theorem I is not stated with reference to tableaux. We wish to point out that although we need Theorem 7 (Chapter V) to prove Theorem I, both Theorem 6 and Theorem 7 of Chapter V can be looked at as special cases of Theorem I (the former by defining $\Gamma(S)$ to mean that all finite subsets of $S$ are satisfiable; the latter by defining $\Gamma(S)$ to mean that no tableau for $S$ can close). Thus the Compactness Theorem for First Order Logic, the SKOLEM-LÖWENHEIM theorem and the Completeness Theorem for tableaux can all be looked at as special cases of our Unifying Principle. And (as we remarked earlier), we shall subsequently see that several other basic results of Quantification Theory can be looked at as special cases of this principle.

In [2] we proved this principle without using analytic tableaux, nor did we use König's lemma (which we used in Chapter V to prove Theorem 7, which in turn we just used to prove Theorem I). The method we used in [2] was to show how a $\Gamma$-consistent set $S$ could be embedded in a

Hintikka sequence $\theta$. The sequence $\theta$ in question is really the leftmost infinite branch of a *systematic* tableau $\mathcal{T}$ for $S$, but it can be generated without consideration of the other branches of $\mathcal{T}$—or without considering trees at all. [The situation is highly analogous to what we did in Chapter III. We first used tableaux to prove the Compactness theorem for Propositional Logic, and then we showed how the essential idea of the construction could be carried out without use of tableaux.]

Here is the construction we used in [2]. We shall now only consider the case when $S$ is *denumerable* (the construction for a *finite* $S$ is even simpler, and the necessary modification should be obvious to the reader.) Suppose $S$ is $\Gamma$-consistent (and pure). Arrange $S$ in some denumerable sequence $X_1, X_2, \ldots, X_n, \ldots$ We take $X_1$ as the first term of $\theta$. This concludes the first stage of the construction. Now suppose we have completed the $n$th stage, and that we have at hand a finite sequence $\theta_n$ of length $\geqslant n$ and that $\theta_n$ is $\Gamma$-consistent with $S$. We then look at the $n$th term $Y$. If it is an $\alpha$, then we extend $\theta_n$ to $\theta_n, \alpha_1, \alpha_2, X_{n+1}$ (and the resulting sequence $\theta_{n+1}$ has at least $n+1$ terms and is $\Gamma$-consistent with $S$ by $A_1$). If $Y$ is a $\beta$, then either $\theta_n, \beta_1, X_{n+1}$ or $\theta_n, \beta_2, X_{n+1}$ is $\Gamma$-consistent with $S$ (by $A_2$), so we let $\theta_{n+1}$ be the former, if $\Gamma$-consistent with $S$, and the latter if otherwise (this really corresponds to going down the *leftmost* $\Gamma$-consistent branch of the systematic tableau). If $Y$ is of the form $\delta$, then we take the first parameter $a$ which is new to $\theta_n$, and we extend $\theta_n$ to $(\theta_n, \delta(a), X_{n+1})$ (which is again $\Gamma$-consistent with $S$ by $A_4$). If $Y$ is of the form $\gamma$, then we take the first parameter $a$ such that $\gamma(a)$ is not a term of $\theta_n$, and we extend $\theta_n$ to $(\theta_n, \gamma(a), \gamma, X_{n+1})$ (which is again $\Gamma$-consistent with $S$ by $A_3$). This concludes stage $n+1$ of the construction. Since each $\theta_n$ is $\Gamma$-consistent with $S$, it cannot contain any element and its conjugate (by $A_0$). Thus the infinite sequence $\theta$ which we generate cannot contain any element and its conjugate. The other four properties of the definition of a Hintikka sequence readily follow from the nature of the construction of $\theta$. So $\theta$ is denumerably satisfiable (by Hintikka's lemma), and every element of $S$ occurs in at least one place in $\theta$. This proves Theorem I.

We see that the above proof nowhere uses König's lemma. So since Theorems 6 and 7 of Chapter V follow from Theorem I, we now see that König's lemma is a convenience rather than a necessity for their proofs.

The above proof of the Unifying Principle is strikingly similar to our third analytic proof of the Compactness Theorem for Propositional Logic given in Chapter III. It is indeed possible to reduce both constructions to a common construction.

There are other ways to prove the Unifying Principle (and hence also the Completeness, Compactness and Skolem-Löwenheim theorems) which make no appeal to the rather careful process for generating an appropriate Hintikka sequence. We shall study this in Chapter X.

## § 3. Analytic Consistency Properties for Finite Sets

For most of the subsequent applications of the Unifying Principle, we will be dealing with properties $\Gamma$ defined just on finite sets. A property $\Gamma$ defined just on finite sets will be called an analytic consistency property (for finite sets) if conditions $A_0 - A_4$ hold for all *finite* sets $S$ (and this definition does *not* require that $\Gamma$ be of finite character). If $\Gamma$ is an analytic consistency property of finite sets, and if a finite set $S$ is $\Gamma$-consistent, then no tableau for $S$ can close, hence $S$ is satisfiable.

**Remarks.** There is of course, a corresponding unifying principle for propositional logic. Define a property $\Gamma$ of sets of formulas of propositional logic to be an analytic consistency property (in the sense of propositional logic) if conditions $A_0, A_1, A_2$, hold. Then it is similarly provable that for such a property $\Gamma$, if a set $S$ has the property $\Gamma$, then $S$ is *truth-functionally* satisfiable. This result simultaneously yields the completeness theorem for propositional tableaux and the Compactness theorem for propositional logic (why?). It can be used to establish the completeness of any of the many well known Hilbert-type axiom systems for propositional logic.

The following exercises anticipate, in part, some results of subsequent chapters.

*Exercise 1.* [A Dual Form of the Unifying Principle]. Let $P$ be a property of finite sets $S$. Call $P$ an *analytic provability property* if the following conditions hold (for every finite set $S$ and every $X, \alpha, \beta, \gamma, \delta$):

$B_0$: $\{S, X, \bar{X}\}$ has the property $P$.

$B_1$: If $\{S, \beta_1\}$ has the property $P$, so does $\{S, \beta\}$; if $\{S, \beta_2\}$ has the property $P$, so does $\{S, \beta\}$.

$B_2$: If $\{S, \alpha_1\}$ and $\{S, \alpha_2\}$ both have the property $P$, so does $\{S, \alpha\}$.

$B_3$: If $\{S, \delta(a)\}$ has the property $P$, so does $\{S, \delta\}$.

$B_4$: If $\{S, \gamma(a)\}$ has the property $P$ and if $a$ does not occur in $\{S, \gamma\}$ then $\{S, \gamma\}$ has the property $P$.

Define a set $S$ to be *disjunctively valid* if for every interpretation I, at least one element of $S$ is true under I. [Note that $S$ is disjunctively valid iff the set $\bar{S}$ of conjugates of elements of $S$ is unsatisfiable.]

(*a*) Show that disjunctive validity is an analytic provability property.

(*b*) Show the following *dual* form of the Unifying Principle: If $P$ is an analytic provability property then every finite set $S$ which is disjunctively valid has the property $P$.

*Exercise 2.* [A Symmetric Form of the Unifying Principle]. Consider a binary relation ⊢ between finite sets—we shall write $U \vdash V$ to mean that the set $U$ stands in the relation $\vdash$ to $V$. Define $\vdash$ to be a *symmetric Gentzen relation* if the following conditions hold (for every $U, V, X, \alpha, \beta, \gamma, \delta$):

$C_0$:  $U, X \vdash V, X$

  $U, X, \bar{X} \vdash V$

  $U \vdash V, X, \bar{X}$

$C_1$: (a) If $U, \alpha_i \vdash V$ then $U, \alpha \vdash V$ $[i = 1, 2]$

  (b) If $U \vdash V, \beta_i$ then $U \vdash V, \beta [i = 1, 2]$

$C_2$: (a) If $U, \beta_1 \vdash V$ and $U, \beta_2 \vdash V$ then $U, \beta \vdash V$

  (b) If $U \vdash V, \alpha_1$ and $U \vdash V, \alpha_2$ then $U \vdash V, \alpha$

$C_3$: (a) If $U, \gamma(a) \vdash V$ then $U, \gamma \vdash V$

  (b) If $U \vdash V, \delta(a)$ then $U \vdash V, \delta$

$C_4$: (a) If $U, \delta(a) \vdash V$ then $U, \delta \vdash V$, provided $a$ does not occur in $U, \delta, V$

  (b) If $U, \vdash V, \gamma(a)$ then $U \vdash V, \gamma$, provided $a$ does not occur in $U, V, \gamma$

Now define $U \vdash_0 V$ to mean that every interpretation which satisfies all elements of $U$ also satisfies *at least* one element of $V$.

(a) Show that $\vdash_0$ is a symmetric Gentzen relation.

(b) Show that for any symmetric Gentzen relation $\vdash$ and for any 2 finite sets $U$, $V$ if $U \vdash_0 V$ then $U \vdash V$.

## Chapter VII

# The Fundamental Theorem of Quantification Theory

There is one very basic result in Quantification Theory which appears to be less widely known and appreciated than it should be. It results from the cumulative efforts of such workers as Herbrand, Gödel, Gentzen, Henkin, Hasenjaeger and Beth. We have referred to it in [2] as a form of Herbrand's theorem, though this is perhaps unfair to the other workers mentioned above. This theorem is indeed Herbrand-like in that it gives a procedure which associates with every valid formula of Quantification Theory a formula of propositional logic which is a *tautology*. This theorem easily yields the completeness theorem for the more conventional axiomatization of First Order Logic (which we study in the next chapter), but it yields far more. The beauty of this theorem is that it makes absolutely no reference to any particular *formal system* of logic; it is stated purely in terms of a certain basic relationship between *first order* satisfiability and *truth-functional* satisfiability. In view of all those considerations, we feel justified in referring to this theorem as the Fundamental Theorem of Quantification Theory.

## § 1. Regular Sets

In preparation for even the statement of the Fundamental Theorem, we must first motivate and then define the notion of a *regular set*.

In this chapter, we will be working only with *unsigned* formulas, and we construe "$\gamma$" and "$\delta$" accordingly. By a *regular* (unsigned) formula we shall mean a *closed* formula which is either of the form $\gamma \supset \gamma(a)$ or of the form $\delta \supset \delta(a)$ where $a$ does *not* occur in $\delta$. We shall refer to regular formulas of the form $\gamma \supset \gamma(a)$ as regular formulas of *type C*, and regular formulas of the form $\delta \supset \delta(a)$ as regular formulas of *type D*.

**Lemma 1.** *Let S be a set of sentences (maybe with parameters) which is (first order) satisfiable. Then*

*(a) For every parameter a, the set $\{S, \gamma \supset \gamma(a)\}$ is satisfiable.*

*(b) For any parameter a which occurs neither in $\delta$ nor any element of S, the set $\{S, \delta \supset \delta(a)\}$ is satisfiable.*

**Proof.** Since $\gamma \supset \gamma(a)$ is *valid*, the proof of (a) is immediate. We turn to the proof of (b).

Suppose $\{S, \delta \supset \delta(a)\}$ were unsatisfiable. Then each of the sets $\{S, \sim \delta\}$ and $\{S, \delta(a)\}$ would be unsatisfiable (because each of the formulas $\sim \delta$ and $\delta(a)$ implies the formula $\delta \supset \delta(a)$). But since $a$ does not occur in $\{S, \delta\}$, the unsatisfiability of the set $\{S, \delta(a)\}$ implies the unsatisfiability of the set $\{S, \delta\}$ (by an argument of the last chapter, if $\{S, \delta\}$ is satisfiable, and $a$ does not occur in $\{S, \delta\}$, then $\{S, \delta, \delta(a)\}$ is satisfiable and hence also the subset $\{S, \delta\}$). Thus both the sets $\{S, \sim \delta\}$ and $\{S, \delta\}$ would be unsatisfiable, hence the set $\{S\}$ would be unsatisfiable, contrary to hypothesis.

**Discussion.** We have proved (b) of Lemma 1 appealing to a fact proved in Chapter V. The following alternative proof proceeds directly (and is very similar to the proof of the corresponding fact in Chapter V).

Suppose $S$ is satisfiable. Let $I$ be any interpretation of all predicates and parameters of $S$ which satisfies $S$. Extend $I$ to any interpretation $I^*$ of the set $\{S, \delta\}$ (i.e. assign to the predicates and parameters of $\delta$ which do not already occur in $S$ any values whatever). Under $I^*$, $\delta$ has a definite truth value, but $\delta(a)$ does not (since we have not yet assigned to the parameter $a$ any value in the universe $\mathsf{U}$ of the interpretation $I^*$). If $\delta$ is false under $I^*$, then we can assign any value of $\mathsf{U}$ to the parameter $a$, and $\delta \supset \delta(a)$ will then be true. If $\delta$ is true under $I^*$, then for some $k \in \mathsf{U}, \delta(k)$ is true under $I^*$, so we take any such $k$ as the value of $a$, and $\delta \supset \delta(a)$ is then true. We have thus extended $I^*$ to an interpretation $I^{**}$ of *all* predicates and parameters of $\{S, \delta \supset \delta(a)\}$ in which all elements of this set are true. Hence $\{S, \delta \supset \delta(a)\}$ is satisfiable.

We shall henceforth use the letter "$Q$" to denote any $\gamma$ or $\delta$, and by $Q(a)$ we respectively mean $\gamma(a)$, $\delta(a)$.

By a *regular sequence* we shall mean a finite (possibly empty) sequence $\langle Q_1 \supset Q_1(a_1), Q_2 \supset Q_2(a_2), ..., Q_n \supset Q_n(a_n) \rangle$ such that every term is regular and such that for each $i < n$, if $Q_{i+1}$ is a $\delta$-formula, then the parameter $a_{i+1}$ does *not* occur in any of the earlier terms $Q_1 \supset Q_1(a_1), ...,$ $Q_i \supset Q_i(a_i)$. By a *regular set $R$* we shall mean a finite set whose members can be arranged in some *regular sequence*. We can alternatively characterize a regular set as any finite set constructed according to the following rules:

$R_0$: The empty set $\emptyset$ is regular.

$R_1$: If $R$ is regular, so is $\{R, \gamma \supset \gamma(a)\}$.

$R_2$: If $R$ is regular, so is $\{R, \delta \supset \delta(a)\}$, *provided $a$ does not occur in $\delta$ nor in $R$.*

Since the empty set is vacuously satisfiable, the conditions $R_1$, $R_2$, and Lemma 1 at once yield (by an obvious induction argument) that every regular set is satisfiable—indeed, if $S$ is any satisfiable set of *pure* sentences and if $R$ is regular, then $S \cup R$ is again satisfiable (for if we successively adjoin the elements of $R$ to $S$, at no stage do we destroy satisfiability (by Lemma 1)). We can say something stronger.

We henceforth shall use "$R$" always to denote a regular set. We shall say that a parameter $a$ occurs *critically* in $R$, or that *$a$ is a critical parameter* of $R$ if there is at least one $\delta$ such that $\delta \supset \delta(a)$ is an element of $R$. Then our induction argument above really yields:

**Theorem 1.** *If $S$ is satisfiable and if no critical parameter of $R$ occurs in $S$, then $R \cup S$ is satisfiable.*

By $\hat{R}$ we shall mean the *conjunction* of the elements of $R$ (in any order and parenthesized in any way).

**Corollary.** *If $\hat{R} \supset X$ is valid, and if no critical parameter of $R$ occurs in $X$, then $X$ is valid. In particular, for any pure sentence $X$, if $\hat{R} \supset X$ is valid, so is $X$.*

**Proof.** Assume hypothesis. Since $\hat{R} \supset X$ is valid, then $\{R, \sim X\}$ is unsatisfiable. Hence $\{\sim X\}$ is unsatisfiable (by Theorem 1, since no critical parameter of $R$ occurs in $\{\sim X\}$), so $X$ is valid.

**Discussion.** Theorem 1, of course, implies that every regular set is satisfiable (just take $S$ in Theorem 1 to be the empty set). Actually regular sets have a property which is stronger than satisfiability. Let us call a set $S$ *sound* if for every interpretation of the predicates of $S$ in a universe U, there exists a choice of values in U for all *parameters* of $S$ which makes all elements of $S$ true. (This notion of *soundness* is, so to speak, intermediate in strength between satisfiability and validity.) Now a regular set is not only satisfiable, but even sound. In fact a regular set $R$ has even the following stronger property: For every interpretation of all predicates of $R$ and all *non-critical* parameters of $R$, there exists a choice of values of the *critical* parameters of $R$ which makes all elements of $R$ true.

*Exercise.* Prove the last statement above.

*Quasi-Regular Sequences.* We will subsequently need:

**Lemma 2.** *Suppose $R$ is regular and that $a$ does not occur in $\delta$ and that $a$ does not occur critically in $R$ (but it may occur non-critically in $R$), and that furthermore no critical parameter of $R$ occurs in $\delta$. Then $\{R, \delta \supset \delta(a)\}$ is regular.*

**Proof.** It is not immediately obvious that we can adjoin $\delta \supset \delta(a)$ to $R$ without destroying regularity (since $a$ might occur in $R$) but we are saved by the other hypothesis. Let $R_1$ be the set of elements of $R$ of type $D$ and arranged in some regular sequence $\delta_1 \supset \delta_1(a_1), \ldots, \delta_n \supset \delta_n(a_n)$, and let $R_2$ be the elements of $R$ of type $C$. By hypothesis, none of the parameters $a_1, \ldots, a_n$ occur in $\delta$, and also each of them is distinct from $a$ (since $a$ does not occur critically in $R$). Therefore, none of the parameters $a_1, \ldots, a_n$ occurs in the formula $\delta \supset \delta(a)$. Hence it follows that the sequence $\langle \delta \supset \delta(a), \delta_1 \supset \delta_1(a_1), \ldots, \delta_n \supset \delta_n(a_n) \rangle$ must be regular. Therefore, $R_1 \cup \{\delta \supset \delta(a)\}$ is a regular set, hence $R_2 \cup R_1 \cup \{\delta \supset \delta(a)\}$ is regular (since all elements of $R_2$ are of type $C$). Thus $\{R, \delta \supset \delta(a)\}$ is regular.

We now define a *quasi-regular* sequence as a sequence $Q_1 \supset Q_1(a_1), \ldots, Q_n \supset Q_n(a_n)$ of regular formulas such that for each $i < n$, either $Q_{i+1}$ is of type $\gamma$ or $a_{i+1}$ does not occur at all in any earlier term $Q_1 \supset Q_1(a_1), \ldots, Q_i \supset Q_i(a_i)$, or else $a_{i+1}$ occurs only *non-critically* in some earlier terms and no critical parameter of any earlier term occurs in $Q_{i+1}$ (and hence also not in $Q_{i+1}(a_{i+1})$). By an obvious induction, Lemma 2 yields:

**Lemma 3.** *Any quasi-regular sequence can be re-arranged to form a regular sequence. Stated otherwise, the set of terms of any quasi-regular sequence is a regular set.*

## § 2. The Fundamental Theorem

We shall first state the following *weak* form of the fundamental theorem.

**Theorem 2.** *Every valid pure sentence $X$ is truth-functionally implied by some regular set $R$. Stated otherwise, for any valid pure sentence $X$, there exists a regular set $R$ such that $\hat{R} \supset X$ is a tautology.*

Theorems 1 and 2 jointly imply that a pure sentence $X$ is valid *if and only if* it is *truth-functionally* implied by some regular set. This is easily seen to be equivalent to the statement that a *finite* set $S$ of pure sentences is unsatisfiable if and only if there exists a regular set $R$ such that $R \cup S$ is *truth-functionally* unsatisfiable.

As we shall later see, Theorem 2 is enough to yield the completeness theorem for the usual axiomatizations of Quantification theory, but for other completeness proofs (which we shall want), we need the full force of the fundamental theorem which is as follows.

Before stating this theorem, we wish to henceforth avoid tiresome use of the phrase "subformula or the negation of a subformula of", and so by a *weak subformula* of $X$ we shall mean a subformula of $X$ or the negation of a subformula of $X$.

**Theorem 2\*** *[The Fundamental Theorem]. Every pure valid sentence $X$ is truth-functionally implied by a regular set $R$ with the additional property that for each member $Q \supset Q(a)$ of $R$, $Q$ is a weak subformula of $X$.*

A set $R$ obeying the conclusion of Theorem 2\* will be called an *associate* of the formula $\sim X$. More generally, for any finite set $S$ of sentences, by an *associate* of $S$ we shall mean a set $R$ satisfying the following 4 conditions:

(1)  $R$ is regular.

(2)  For each member $Q \supset Q(a)$ of $R$, $Q$ is a weak subformula' of some element of $S$.

(3)  No *critical* parameter of $R$ occurs in $S$.

(4)  $R \cup S$ is truth-functionally unsatisfiable.

The fundamental theorem can be equivalently stated thus: Every unsatisfiable pure finite set $S$ has an associate. [We also note that if $S$ has an associate, then $S$ is unsatisfiable by Theorem 1.]

We shall give several proofs of the fundamental theorem in the course of this study. We shall now establish it as a consequence of our Unifying Principle.

Define $\Gamma(S)$ to mean that $S$ has no associate—or put otherwise, call $S$ $\Gamma$-*inconsistent* if $S$ *has* an associate. It only remains to verify that this $\Gamma$ is an analytic consistency property. It will be somewhat more convenient to use the definition of "analytic consistency property" in the form of the conditions $A_0' - A_4'$ rather than $A_0 - A_4$ (cf. § 1 of the preceding chapter). So we verify $A_0' - A_4'$, reading "$S$ has an associate" for "$S$ is $\Gamma$-inconsistent".

(0) – If $S$ contains some atomic element and its negation, then $S$ is already *truth-functionally* unsatisfiable, in which case the empty set $\emptyset$ is an associate of $S$.

(1) – If $\{S, \alpha, \alpha_1\}$ or $\{S, \alpha, \alpha_2\}$ has an associate $R$, then $R$ is also an associate of $\{S, \alpha\}$.

(2) – Suppose $R_1$ is an associate of $\{S, \beta_1\}$ and $R_2$ is an associate of $\{S, \beta_2\}$. If no critical parameter of $R_1$ occurs in $R_2$ or $\beta_2$, and no critical

parameter of $R_2$ occurs in $R_1$ or $\beta_1$, then $R_1 \cup R_2$ is regular and is an associate of $\{S,\beta\}$ (verify!). If there are critical parameters $a_1, \ldots, a_n$ of $R_1$ which also occur in $R_2$, then we take parameters $b_1, \ldots, b_n$ which occur nowhere in $R_1 \cup R_2 \cup \{S,\beta\}$ and we everywhere replace $a_1, \ldots, a_n$ by $b_1, \ldots, b_n$ respectively in $R_1$, and let $R_1^*$ be the resulting set. Then $R_1^*$ is obviously also an associate of $\{S,\beta_1\}$ and has no critical parameter which occurs in $R_2$. Then $R_1^* \cup R_2$ is an associate of $\{S,\beta\}$.

(3)—If $R$ is an associate of $\{S,\gamma,\gamma(a)\}$ then $\{R,\gamma \supset \gamma(a)\}$ is an associate of $\{S,\gamma\}$.

To see this, we first note the obvious fact that $\{R,\gamma \supset \gamma(a)\}$ is regular. Secondly, $\gamma$ is a subformula of some element of $\{S,\gamma\}$ )viz. $\gamma$), and for every $Q \supset Q(b)$ in $R$, $Q$ is a weak subformula of (some element of) the set $\{S,\gamma,\gamma(a)\}$, and if it is a weak subformula of $\gamma(a)$, it is also a weak subformula of $\gamma$). Thirdly, the critical parameters of $\{R,\gamma \supset \gamma(a)\}$ are the same as those of $R$. Since $R$ is an associate of $\{S,\gamma,\gamma(a)\}$, none of them occur in $\{S,\gamma,\gamma(a)\}$, so none of them occur in $\{S,\gamma\}$). Fourthly, the set $\{R, \gamma \supset \gamma(a),S,\gamma\}$ must be truth-functionally unsatisfiable, because the conjunction of its elements *truth-functionally* implies that all elements of $\{R,S,\gamma,\gamma(a)\}$ are true, but this latter set is by hypothesis not truth-functionally satisfiable, hence the former set is truth-functionally unsatisfiable.

(4) – Suppose $R$ is an associate of $\{S,\delta(a)\}$ and that $a$ does not occur in $S$ nor $\delta$, then $\{R,\delta \supset \delta(a)\}$ is an associate of $\{S,\delta\}$.[1]

To see this, the delicate point to verify is that the set $\{R,\delta \supset \delta(a)\}$ is actually regular. But this follows from Lemma 2 of § 1, since our present hypothesis says that $a$ does not occur in $\delta$ nor *critically* in $R$ (since no parameter of $\{S,\delta(a)\}$ does) nor does any critical parameter of $R$ occur in $\delta$ (because it does not occur in $\{S,\delta(a)\}$, hence also not in $\delta(a)$, and hence also not in $\delta$). Therefore, the hypothesis of Lemma 2 holds, so $\{R,\delta \supset \delta(a)\}$ is regular. The verification of the remaining 3 points necessary to show that $\{R,\delta \supset \delta(a)\}$ is an associate of $\{S,\delta\}$ is essentially the same as in (3) (though the one new feature in the verification that no critical parameter of $\{R,\delta \supset \delta(a)\}$ occurs in $\{S,\delta\}$ is the fact that $a$, though a critical parameter of $\{R,\delta \supset \delta(a)\}$ does not occur in $\{S,\delta\}$ by hypothesis). This concludes the proof.

## § 3. Analytic Tableaux and Regular Sets

Suppose we have a closed tableau $\mathcal{T}$ for a finite set $S$ of pure sentences. Then $S$ must be unsatisfiable, hence by the Fundamental Theorem, $S$ must have an associate $R$. Now, given the tableau $\mathcal{T}$, how can we

---

[1] Assuming $a$ occurs in $\delta(a)$ (otherwise $R$ itself is an associate of $\{S, \delta\}$).

*effectively* find such an $R$? The answer is delightfully simple—just take for $R$ the set of all formulas $Q \supset Q(a)$ such that $Q(a)$ was inferred from $Q$ (by Rule $C$ or $D$) on the tableau $\mathscr{T}$! We easily show that this $R$ is an associate of $S$ (and without use of the Fundamental Theorem, so we also have another proof of the Fundamental Theorem using the Completeness theorem for tableaux).

That $R$ is regular is immediate from the restriction in Rule $D$. Also, for any element $Q \supset Q(a)$ of $R$, the formula $Q$ occurs as a point on the tableau $\mathscr{T}$, hence must be a weak subformula of some element of $S$. Since $S$ contains no parameters, then of course, no critical parameter of $R$ occurs in $S$. It thus remains to show that $S \cup R$ is *truth-functionally* unsatisfiable. We shall do this by showing how to construct a closed tableau $\mathscr{T}_0$ for the set $S \cup R$ which uses only the truth-functional rules (Rules $A$ and $B$).

Before doing this, we pause to discuss a point (which perhaps we should have discussed earlier) concerning the use of *Modus Ponens* within tableaux. The rule of *modus ponens*—a common rule in most logical systems—says "From $X$ and $X \supset Y$ to infer $Y$". Suppose we add this to our tableaux rules in the form "given a branch $\theta$ of $\mathscr{T}$ containing $X$ and $X \supset Y$ as terms, we may adjoin $Y$ as an end point to $\theta$—i.e., we may extend $\theta$ to $\theta, Y$". Does this additional rule increase the class of provable formulas? The answer is quite trivially "no", for given a branch $\theta$ containing $X$ and $X \supset Y$, since it contains $X \supset Y$, then we can (by Rule $B$) simultaneously extend it to the two branches $\theta, \sim X$ and $\theta, Y$, but the left extended branch $\theta, \sim X$ is immediately closed (since $\theta$ contains $X$), so we have, in effect, extended $\theta$ to $\theta, Y$.

Looked at diagramatically, whereas by modus ponens we do this:

$$\theta \left\{ \begin{array}{l} \vdots \\ X \\ \vdots \\ X \supset Y \\ \vdots \\ Y \end{array} \right.$$

without modus ponens we do this:

$$\theta \left\{ \begin{array}{l} \vdots \\ X \\ \vdots \\ X \supset Y \end{array} \right.$$
$$\frac{\sim X \mid Y}{X}$$

Let us now return to the problem of showing $R \cup S$ to be truth-functionally unsatisfiable by constructing the tableau $\mathcal{T}_0$ for $R \cup S$ using only Rules $A$, $B$.

We simply change $\mathcal{T}$ to $\mathcal{T}_0$ as follows. Whereas in $\mathcal{T}$ we could freely infer $Q(a)$ from $Q$ by a quantificational rule, in $\mathcal{T}_0$, when we have a branch $\theta_1, Q, \theta_2$ and we wish to infer $Q(a)$, we first adjoin $Q \supset Q(a)$ (as a premise of the tableau, since it occurs in $R$ and hence in $R \cup S$) and then infer $Q(a)$ by modus ponens. Looked at diagrammatically, in $\mathcal{T}$ we have the following:

$$\left.\begin{array}{c} \vdots \\ Q \\ \vdots \end{array}\right\} \theta$$

$$Q(a) \quad [\text{by Rule } C \text{ or } D]$$

We alter this in $\mathcal{T}_0$ as follows:

$$\left.\begin{array}{c} \vdots \\ Q \\ \vdots \end{array}\right\} \theta$$

$$Q \supset Q(a)$$
$$Q(a)$$

Or more completely (without use of Modus Ponens):

$$\left.\begin{array}{c} \vdots \\ Q \\ \vdots \end{array}\right\} \theta$$

$$Q \supset Q(a) \quad [\text{element of } R]$$
$$\underline{\qquad\qquad\qquad}$$
$$\sim Q \mid Q(a)$$
$$X$$

Thus the set $R \cup S$ is *truth-functionally* unsatisfiable.

If $X$ is a valid pure sentence and if $R$ is an associate of $\{\sim X\}$, let us refer to the tautology $\hat{R} \supset X$ as an *associated tautology* of $X$. It is instructive and often curious to see what the associated tautology looks like. We have just given a procedure whereby we can find an associated tautology for $X$ given a closed tableau for $\{\sim X\}$. Let us consider an example. The formula $(\exists x)[Px \vee Qx] \supset [(\exists x)Px \vee (\exists x)Qx]$—which we will call $A$—is valid; the following is a closed tableau for its negation:

(1) $\sim[(\exists x)(Px \vee Qx) \supset ((\exists x)Px \vee (\exists x)Qx)]$
(2) $(\exists x)(Px \vee Qx)$      (1)
(3) $\sim((\exists x)Px \vee (\exists x)Qx)$      (1)
(4) $\sim(\exists x)Px$      (3)
(5) $\sim(\exists x)Qx$      (3)
(6) $Pa \vee Qa$      (2)
(7) $\sim Pa$      (4)
(8) $\sim Qa$      (5)
(9) $Pa$ (6) $\mid$ (10) $Qa$ (6)

We have used the quantification rules $(C \& D)$ to infer (6) from (2), (7) from (4) and (8) from (5). Accordingly, our derived $R$ is the set:

$$\left\{\begin{array}{c} (2) \supset (6) \\ (4) \supset (7) \\ (5) \supset (8) \end{array}\right\}$$

that is,

$$R = \left\{\begin{array}{c} (\exists x)(Px \vee Qx) \supset (Pa \vee Qa) \\ \sim(\exists x)Px \supset \sim Pa \\ \sim(\exists x)Qx \supset \sim Qa \end{array}\right\}$$

According to our theory, it should truth functionally imply

$$A = (\exists x)(Px \vee Qx) \supset ((\exists x)Px \vee (\exists x)Qx).$$

To see this more perspicuously, let us abbreviate all *Boolean atoms* involved by propositional variables. We let

$$\begin{array}{l} p_1 = (\exists x)(Px \vee Qx) \\ p_2 = Pa \\ p_3 = Qa \\ p_4 = (\exists x)Px \\ p_5 = (\exists x)Qx \end{array}$$

then we have

$$R = \left\{\begin{array}{c} p_1 \supset (p_2 \vee p_3) \\ \sim p_4 \supset \sim p_2 \\ \sim p_5 \supset \sim p_3 \end{array}\right\}$$

$$A = p_1 \supset (p_4 \vee p_5).$$

Without even knowing what "$p_1$", ..., "$p_5$" stand for, the reader can see that $\dot{R} \supset A$ is a tautology.

## § 4. The Liberalized Rule $D$

Now we are in a good position to justify the soundness of the liberalized version of Rule $D$ for tableaux. We shall refer to this liberalized rule as Rule $D^*$, and we recall that it says "from $\delta$ we may infer $\delta(a)$ provided that either $a$ is new, or else the following 3 conditions all hold: (1) $a$ does not occur in $\delta$; (2) $a$ has not been previously introduced by Rule $D^*$; (3) no parameter which has been previously introduced by Rule $D^*$ occurs in $\delta$".

Suppose $\mathscr{T}$ is a tableau for $S$. Again let $R$ be the set of all formulas $Q \supset Q(a)$ such that $Q(a)$ was inferred from $Q$ on the tableau $\mathscr{T}$, and let $\dot{R}$ be the *sequence* $\langle Q_1 \supset Q_1(a_1), ..., Q_n \supset Q_n(a_n) \rangle$ of elements of $R$

arranged in the order in which the rules $C$ and $D$ were applied in the tableaux (i.e. in $\mathcal{T}$ we first inferred $Q_1(a_1)$ from $Q_1, Q_2(a_2)$ from $Q_2, ..., Q_n(a_n)$ from $Q_n$). Now if $\mathcal{T}$ was constructed using the strict Rule $D$, then not only is the set $R$ regular, but the *sequence* $\dot{R}$ is a regular sequence. If, however, $\mathcal{T}$ was constructed using the liberalized Rule $D^*$, then the *sequence* $\dot{R}$ is not necessarily regular, but it must be *quasi-regular*. Therefore the *set* $R$ is still a regular set (by Lemma 3). Furthermore $R \cup S$ is truth-functionally unsatisfiable (by the same argument we used for tableaux constructed according to the strict Rule $D$—i.e., we can construct a closed tableau for $R \cup S$ using only Rules $A$, $B$). If now $S$ is a *pure* set then $S$ is unsatisfiable (by theorem 1).

Let us call a tableau $\mathcal{T}$ constructed using Rule $D^*$ a *liberalized* tableau and let us refer to those parameters of $\mathcal{T}$ which were introduced by Rule $D^*$ as the *critical* parameters of $\mathcal{T}$—we also say that they occur or were introduced *critically* in $\mathcal{T}$.

We have so far considered only tableaux for *pure* sets $S$. However, our above argument shows the following (where now $S$ may contain parameters).

**Theorem.** (*a*) *If there exists a closed liberalized tableau for $S$, and if no parameter of $S$ was introduced critically in the tableau, then $S$ is unsatisfiable.*

(*b*) – *If $X$ is provable by a liberalized tableau, and if no parameter of $X$ occurs critically in the tableau, then $X$ is valid.*

<br>

## Chapter VIII

# Axiom Systems for Quantification Theory

We now wish to show how we can use our earlier completeness results to establish the completeness of the more usual formalizations of First Order Logic. We first consider an axiom system $Q_1$ which (except for rather minor details) is a standard axiom system. Its completeness is easily obtained as a consequence of our Unifying Principle. We prefer, however, to emphasize a completeness proof along the following lines. We consider a succession of axiom systems, starting with $Q_1$ and ending with a system $Q_2^*$ each of which seems ostensibly weaker than the preceding. The completeness of $Q_2^*$ easily implies the completeness of $Q_1$, but it is not immediately obvious that everything provable in $Q_1$ is provable in $Q_2^*$. However, the completeness of $Q_2^*$ is an almost immediate consequence of the (strong form of) the Fundamental Theorem. The arguments of this chapter are wholly constructive. We give an effective procedure whereby given any associate of $\{\sim X\}$, a proof of $X$ can be found in $Q_2^*$ (and hence also in $Q_1$). We also know from the preceding chapter how given a closed tableau for $\{\sim X\}$, we can find an associate of $\{\sim X\}$. Combining these two constructions, we see how given any proof of $X$ by the tableau method, we can find a proof of $X$ in each of the axiom systems of this chapter.

## § 0. Foreword on Axiom Systems

By an axiom system $\mathscr{A}$ is meant a domain $D$ of elements called *formal objects* together with a subset $A$ of $D$ whose elements are called the *axioms* of the system together with a set of relations (of elements of $D$) called *rules of inference* or *inference rules*. If $R$ is an inference rule and if $R(X_1,...,X_n, Y)$ holds, then we say that $\langle X_1,...,X_n, Y \rangle$ is an *application* of the rule $R$, and that $Y$ is a *direct consequence* of $X_1,...,X_n$ under $R$, or that $Y$ is directly derivable from $X_1,...,X_n$ under $R$. In any application $\langle X_1,...,X_n, Y \rangle$ of $R$, the elements $X_1,...,X_n$ are called the *premises* of the application and $Y$ is called the *conclusion* of the application. By a *proof* in $\mathscr{A}$ is meant a finite sequence $X_1,...,X_n$ such that each term $X_i$ is either an axiom of $\mathscr{A}$ or is directly derivable from one or more earlier terms of the sequence under one of the inference rules of $\mathscr{A}$. A proof $Y_1, ..., Y_n$ in $\mathscr{A}$ is also called a proof of its last term $Y_n$ and finally an element $X$ is called *provable in $\mathscr{A}$* or a *theorem* of $\mathscr{A}$ if there exists a proof of $X$ in $\mathscr{A}$.

In the so-called *Hilbert-type* axiom systems for propositional logic (quantification theory) the formal objects are formulas of propositional logic (respectively quantification theory.) In this chapter we consider only Hilbert-type systems. [In chapter XI we shall consider the so-called *Gentzen-type* systems in which the formal objects are of a slightly more complex nature.]

Following Kleene [1] we use the term "postulate" collectively for axioms and inference rules. We do not require that the set of axioms be finite. One sometimes displays an infinite set of axioms by means of a so-called *axiom scheme* which specifies the set of all expressions of such and such form. [An example of a typical axiom scheme for propositional logic is the set of all formulas of the form $(X \wedge Y) \supset X$. An example of an axiom scheme for quantification theory—one which we will use in our system $Q_1$—is the set of all sentences of the form $(\forall x)X \supset X_a^x$.]

Inference rules being usually displayed in the form of a figure in which a horizontal line is drawn, the premises are written above the line and the conclusion below the line. For example, the rule of modus ponens is displayed thus:

$$\frac{X, X \supset Y}{Y}$$

It is read "$Y$ is directly derivable from the two premises $X, X \supset Y$" or "from $X, X \supset Y$".

Many axiom systems for propositional logic exist in the literature in which modus ponens is the only rule of inference. In the older axiomatic treatments of quantification theory, it was customary to give axioms for propositional logic in addition to axioms for the quantifiers. In more modern versions, one often takes all *tautologies* as axioms (in addition to the quantification axioms). This is the course we shall adopt, since an axiomatic analysis of logic (along Hilbert-type lines) at the propositional level is not needed for any purposes considered in this study. [Later, though, we will consider Gentzen-type axiom systems at the propositional level, and these will be useful.]

## § 1. The System $Q_1$

In displaying the axiom schemata and inference rules of this system, we shall use the following notation: $\varphi(x)$ is to be any formula which has no free variable other than $x$, and $\varphi(a)$ is to be the result of substituting the parameter $a$ for every free occurrence of $x$ in $\varphi(x)$.

*Axiom Schemata:*    I.   All closed tautologies

                II.   $(\forall x)\varphi(x) \supset \varphi(a), \quad \varphi(a) \supset (\exists x)\varphi(x)$

*Inference Rules:*    I.   (Modus Ponens) $\dfrac{X \quad X \supset Y}{Y}$

                II.   (Generalization Rules)

$$\frac{X \supset \varphi(a)}{X \supset (\forall x)\varphi(x)} \qquad \frac{\varphi(a) \supset X}{(\exists x)\varphi(x) \supset X}$$

provided $a$ does not occur in $X$ nor in $\varphi(x)$.

**Remarks.** The reader might find it of interest to compare the system $Q_1$ with say the systems given in Church [1] or Kleene [1]. To the reader familiar with these systems, we wish to point out that in our version $Q_1$, we have arranged matters so that we completely avoid the bothersome trouble of collision of quantifiers. Roughly speaking, we have done this by weakening Axiom Scheme II and strengthening the Generalization Rules.

*Exercise 1.* Show that the above generalization rules *preserve validity* —i.e. if $X \supset \varphi(a)$ is valid then $X \supset (\forall x)\varphi(x)$ is valid, and if $\varphi(a) \supset X$ is valid, then $(\exists x)\varphi(x) \supset X$ is valid. It is, of course, obvious that all axioms of $Q_1$ are valid, and that modus ponens preserves validity—i.e. if $X$ and $X \supset Y$ are both valid, then $Y$ is valid. Show therefore that the system $Q_1$ is *correct* in the sense that everything provable in $Q_1$ is actually valid. Show that $Q_1$ is *consistent* in the sense that no formula and its negation are both provable in $Q_1$.

*Exercise 2.* Give the following *finitary* proof of the consistency of $Q_1$: Show that all axioms of $Q_1$ are valid in a 1-element domain, and that the inference rules all preserve validity in a 1-element domain. Show therefore, that every theorem of $Q_1$ is valid in a 1-element domain. Show therefore that $Q_1$ is consistent.

Before proceeding further we wish to point out that *any* axiom system which contains all tautologies as axioms and which has modus ponens as one of its inference rules, must be closed under *truth-functional implication*—i.e. if $X_1, ..., X_n$ are provable in the system and if $Y$ is truth-functionally implied by $\{X_1, ..., X_n\}$, then $Y$ is provable in the system. For suppose the system does contain all tautologies as axioms (or for that matter, all tautologies as theorems) and that the system has modus ponens as an inference rule (or for that matter, we only require that the systems be *closed* under modus ponens—i.e., for every $X$ and $Y$, if $X$, $X \supset Y$ are both theorems, so is $Y$). Now suppose that $X_1, ..., X_n$ are theorems of the system and that $Y$ is truth-functionally implied by $\{X_1, ..., X_n\}$. Then $X_1 \supset (X_2 \supset (\cdots \supset (X_n \supset Y)...)$ is a tautology, hence is a theorem of the system. Since this and $X_1$ are both theorems, so is $X_2 \supset (\cdots \supset (X_n \supset Y)...)$ (by closure under modus ponens). Since this and $X_2$ are both theorems, so is $X_3 \supset (\cdots \supset (X_n \supset Y)...)$. After $n$ rounds of this reasoning, we see that $Y$ must be a theorem of the system. Let us record this fact as

**Lemma 1.** *Any axiom system which contains all tautologies as theorems and which is closed under modus ponens must be closed under truth-functional implication.*

We next wish to consider a trivial variant $Q_1'$ of $Q_1$ which enables us to handle the quantifiers in the unified "$\gamma, \delta$" manner. To obtain $Q_1'$ we replace the right half of Axiom Scheme II (viz. $\varphi(a) \supset \exists x \varphi(x)$ by $\sim(\exists x)\varphi(x) \supset \sim\varphi(a)$, and we replace the left half of the Generalization Rule (viz. "from $X \supset \varphi(a)$ to infer $X \supset (\forall x)\varphi(x)$ (with proviso)" by "from $\sim\varphi(a) \supset X$ to infer $\sim(\forall x)\varphi(x) \supset X$ (with same proviso)". Thus the postulates of $Q_1'$ are as follows:

I′: All closed tautologies

II′: Modus Ponens

III′: $(\forall x)\varphi(x) \supset \varphi(a)$ $\qquad \sim(\exists x)\varphi(x) \supset \sim\varphi(a)$

IV′: $\dfrac{\sim\varphi(a) \supset X}{\sim(\forall x)\varphi(x) \supset X}$ $\qquad \dfrac{\varphi(a) \supset X}{(\exists x)\varphi(x) \supset X}$

(with proviso)

We can now rewrite III′ and IV′ in unified notation as follows:

III′: $\gamma \supset \gamma(a)$

IV′: $\dfrac{\delta(a) \supset X}{\delta \supset X}$, provided $a$ does not occur in $\delta$ nor in $X$.

We leave to the reader the trivial verification (using Lemma 1) that the systems $Q_1, Q_1'$ are equivalent (in the sense that a formula is provable in the one system iff it is provable in the other).

Next we consider a variant $Q_1''$ of $Q_1'$ which allows a certain useful symmetry. The postulates of $Q_1''$ are:

I″: All closed tautologies

II″: Modus Ponens

III″$_a$: $\dfrac{\gamma(a) \supset X}{\gamma \supset X}$

III″$_b$: $\dfrac{\delta(a) \supset X}{\delta \supset X}$ providing $a$ does not occur in $\delta$ nor in $X$.

Thus the only *axioms* of $Q_1''$ are the closed tautologies — the other postulates are all inference rules. The sole difference between $Q_1'$ and $Q_1''$ is that $Q_1''$ contains the inference rule III″$_a$ in place of the *axiom scheme* III′ of the system $Q_1'$.

The equivalence of the systems $Q_1'$ and $Q_1''$ is easily seen as follows. First of all $\gamma(a) \supset (\gamma \supset \gamma(a))$ is a tautology, hence is a theorem (indeed an axiom) of $Q_1''$. Therefore by III″$_a$ (taking $\gamma \supset \gamma(a)$ for $X$), $\gamma \supset (\gamma \supset \gamma(a))$ is a theorem of $Q_1''$. But $\gamma \supset \gamma(a)$ is truth-functionally implied by $\gamma \supset (\gamma \supset \gamma(a))$, hence $\gamma \supset \gamma(a)$ is a theorem of $Q_1''$ (by Lemma 1). From this it is immediate that everything provable in $Q_1'$ is provable in $Q_1''$.

To show that everything provable in $Q_1''$ is provable in $Q_1'$, we need merely show that Rule III″$_a$ is *derivable* in $Q_1'$ in the sense that in any application of the rule, if the premise is provable in $Q_1'$, so is the conclusion. Thus we must show that if $\gamma(a) \supset X$ is provable in $Q_1'$, so is $X$. So suppose $\gamma(a) \supset X$ is provable in $Q_1'$. Obviously $\gamma \supset \gamma(a)$ is also provable in $Q_1'$ (it is an axiom). Hence $\gamma \supset X$ is provable in $Q_1'$ by Lemma 1, since $\gamma \supset X$ is truth-functionally implied by $\{\gamma \supset \gamma(a), \gamma(a) \supset X\}$. We thus see that the systems $Q_1, Q_1', Q_1''$ are quite trivially equivalent.

It will be convenient to unite rules $\text{III}''_a$ and $\text{III}''_b$ as follows:

$\text{III}''$: $\dfrac{Q(a) \supset X}{Q \supset X}$, where $Q$ is of type $D$ and $a$ does not occur in $Q \supset X$, or $Q$ is of type $C$.

## § 2. The Systems $Q_2$, $Q_2^*$

We now turn to a system which in some ways is more interesting. The postulates of $Q_2$ are as follows:

$A_1$: All closed tautologies

$B_1$: $\dfrac{[\gamma \supset \gamma(a)] \supset X}{X}$

$B_2$: $\dfrac{[\delta \supset \delta(a)] \supset X}{X}$, provided $a$ does not occur in $\delta$ nor in $X$.

We emphasize that modus ponens is *not* an inference rule of $Q_2$—indeed both rules $B_1$, $B_2$ are *one-premise* rules. Thus it is not immediately obvious that the set of theorems of $Q_2$ is closed under truth-functional implication.

It will be convenient to rewrite our postulates as follows:

$A$: All closed tautologies

$B$: $\dfrac{[Q \supset Q(a)] \supset X}{X}$, provided $Q$ is of type $D$ and $a$ does not occur in $Q$ nor $X$, or $Q$ is of type $C$.

**Theorem 1.** *Everything provable in $Q_2$ is provable in $Q_1''$ (and hence also in $Q_1'$ and in $Q_1$).*

**Proof.** It suffices to show that $B$ is a derived rule of $Q_1''$—i.e. that in any application, if the premise is provable in $Q_1''$, so is the conclusion. So suppose that $[Q \supset Q(a)] \supset X$ is provable in $Q_1''$ and that the proviso of Rule $B$ holds. Then the proviso of Rule $\text{III}''$ of $Q_1''$ also holds. Since $[Q \supset Q(a)] \supset X$ is provable in $Q_1''$, so are both $\sim Q \supset X$, $Q(a) \supset X$ (by Lemma 1). Since $Q(a) \supset X$ is provable in $Q_1''$ and $a$ does not occur in $Q$ nor in $X$ if $Q$ is of type $D$, then $Q \supset X$ is provable in $Q_1''$ (by Rule $\text{III}''$). Thus $\sim Q \supset X$ and $Q \supset X$ are both provable in $Q_1''$, hence (by Lemma 1) $X$ is provable in $Q_1''$. This concludes the proof.

Next we consider the system $Q_2^*$ whose postulates are as follows:

$A^*$: Same as $A$.

$B^*$: Like $B$ with the additional proviso that $Q$ be a weak subformula of $X$.

Of course $Q_2^*$ is immediately a subsystem of $Q_2$ in the strong sense that any proof in $Q_2^*$ is already a proof in $Q_2$.

**Theorem 2.** *Given an associate of* $\sim X$, *a proof of* $X$ *in* $Q_2^*$ *can be found.*

**Proof.** Let $R$ be an associate of $\sim X$. If $R$ is empty, then $X$ is a tautology, hence we immediately have a proof of $X$ in $Q_2^*$. If $R$ is non-empty, then we arrange $R$ in an *inverse* regular sequence $Y_1, \ldots, Y_n$, (i.e. each $Y_i$ is of the form $Q_i \supset Q_i(a_i)$, where $Q_i$ is a weak subformula of $X$ and, if of type $D$, then $a_i$ does not occur in any of $Y_{i+1}, \ldots, Y_n, X$). Also $Y_1 \supset (Y_2 \supset \cdots \supset (Y_n \supset X)\ldots)$ is a tautology, hence is a theorem (indeed an axiom) of $Q_2^*$. Then by our sole inference rule of $Q_2^*$, $Y_2 \supset \cdots \supset (Y_n \supset X)\ldots)$ is a theorem of $Q_2^*$. If $n > 1$, then another application of the inference rule (B) gives a proof of $Y_3 \supset (\cdots \supset (Y_n \supset X)\ldots)$. In this manner we successively eliminate $Y_1, \ldots, Y_n$ and obtain a proof of $X$. In other words, the following sequence of lines is a proof of $X$ in $Q_2^*$:

$$Y_1 \supset (Y_2 \supset \cdots \supset (Y_n \supset X)\ldots)$$
$$Y_2 \supset (\cdots \supset (Y_n \supset X)\ldots)$$
$$\vdots$$
$$Y_n \supset X$$
$$X$$

This concludes the proof.

Now Theorem 2 and the Fundamental Theorem of Quantification Theory at once yields.

**Theorem 3.** *(A strong completeness theorem). Every valid sentence is provable in* $Q_2^*$.

Now by Theorems 1 and 3 (and the fact that $Q_2^*$ is a subsystem of $Q_2$) we have

**Theorem 4.** *(Gödel's Completeness Theorem)*[1]. *Every valid sentence is provable in* $Q_1$.

**Discussion.** The completeness of $Q_2^*$ is a better result than the completeness of $Q_1$—it is closely related to the work of Herbrand. The system $Q_2^*$ has a feature reminiscent of Gentzen's Extended Hauptsatz (which we shall subsequently study)—viz. that any sentence provable in $Q_2^*$ can be proved by first using truth-functional rules and then quantificational ones. (Indeed this is the *only* way a sentence can be proved in this particular system $Q_2^*$!)

---

[1] Proved by Gödel for a slightly different system than $Q_1$.

There is a way of modifying the system $Q_2^*$, which preserves the above mentioned feature, and has the additional feature that many proofs use only weak subformulas of the sentence to be proved. This can be done as follows. For any $Q$ of the form $(\forall x)X, (\exists x)X$, define $\bar{Q}$ to be $\sim(\forall x)X$, $\sim(\exists x)X$ respectively. And for any $Q$ of the form $\sim(\forall x)X, \sim(\exists x)X$, define $\bar{Q}$ to be $(\forall x)X, (\exists x)X$ respectively. Now let $Q_2^{**}$ be $Q_2^*$ with rule $B$ replaced by the following (2-premise) rule:

$$\frac{\bar{Q} \supset X \quad Q(a) \supset X}{X}, \text{ provided } \bar{Q} \text{ is a weak subformula of } X, \text{ and}$$

that if $Q$ is of the $\delta$-type, then $a$ does
not occur in $Q$ nor $X$.

The reader should find it a profitable exercise to prove the completeness of $Q_2^{**}$. The reader is also urged to look at the following exercises:

*Exercise 1.* Define a finite set $\{X_1, ..., X_n\}$ to be *refutable* or *inconsistent* in $Q_1(Q_1', Q_1'')$ if the sentence $\sim(...(X_1 \wedge X_2) \wedge \cdots \wedge X_n)$ is provable in $Q_1(Q_1', Q_1'')$. Establish the completeness of $Q_1(Q_1', Q_1'')$ as a corollary of the Unifying Principle (Chapter VI) by showing directly that consistency in $Q_1(Q_1', Q_1'')$ is an *analytic consistency property* (for finite sets).

*Exercise 2.* Modify the above argument to give an alternative (and possibly more direct) process for translating a proof by analytic tableaux to a proof in $Q_1(Q_1', Q_1'')$. More specifically, call a tableau $\mathscr{T}$ (for *unsigned* formulas) refutable in $Q_1(Q_1', Q_1'')$ if (the set of elements of) each branch of $\mathscr{T}$ is refutable in $Q_1(Q_1', Q_1'')$. Obviously a closed tableau is refutable in $Q_1$ (why?). Now suppose $\mathscr{T}'$ is an immediate extension of $\mathscr{T}$ (by one application of Rule $A$, $B$, $C$, or $D$)—show how a refutation of $\mathscr{T}'$ in $Q_1(Q_1', Q_1'')$ leads to a refutation of $\mathscr{T}$ in $Q_1(Q_1', Q_1'')$. Then by induction, show the *origin* of any closed tableau can be refuted in $Q_1(Q_1', Q_1'')$.

Chapter IX

# Magic Sets

There is another approach to the Completeness, Compactness and Skolem-Löwenheim theorems along very different lines which is most striking in its simplicity. It is an outgrowth of the completeness proofs of Henkin and Hasenjaeger and we shall study its relation to those methods in the next chapter. There are certain key sets involved which have almost "magical" properties, and which we accordingly call magic sets.[1]) [They are closely related to the methods of Hilbert's $\varepsilon$-calculus.]

---

[1]) The use of the term "magic" in mathematics is not totally new; one speaks, e. g. of "magic squares".

## § 1. Magic Sets

In this chapter we will be working exclusively with *unsigned* formulas. Also Boolean valuations and First Order valuations will always be understood as referring to valuations of the set $E^V$ of all closed formulas using only denumerably many parameters.

**Definition.** By a *magic set* we shall mean a set $M$ of sentences with (or without) parameters such that:

$M_1$: Every Boolean valuation satisfying $M$ is also a First Order valuation.

$M_2$: For any *finite* set $S_0$ of *pure* sentences and for every *finite* subset $M_0$ of $M$, if $S_0$ is (first-order) satisfiable, then so is $S_0 \cup M_0$.

The first surprising fact about magic sets is that they exist (which we will subsequently show). One importance of magic sets emerges from the following considerations.

We shall say that a formula $X$ is *truth-functionally deducible* from a set $S$ if there exists a *finite* subset $S_0$ of $S$ such that $X$ is *truth-functionally* implied by $S_0$ (or stated otherwise, that the formula $\hat{S}_0 \supset X$ is a *tautology*). We shall say that a set $B$ constitutes a *truth-functional basis for Quantification Theory* if for every *pure* sentence $X$, $X$ is valid if and only if $X$ is truth-functionally deducible from $B$. Now our first theorem is

**Theorem 1.** *Every magic set $M$ is a truth-functional basis for Quantification Theory.*

**Proof.** Let $M$ be a magic set.

(a) Suppose $X$ is valid. We assert that $X$ must be true in all *Boolean* valuations which satisfy $M$. For consider any Boolean valuation $v$ which satisfies $M$. Then $v$ is also a First Order valuation (by $M_1$). Hence $X$ must be true under $v$ (because a valid formula, by definition, is true under all First Order valuations).

Thus $X$ is indeed true under all Boolean valuations which satisfy $M$. Then by the Compactness theorem for *propositional* logic (in the second form for *deducibility*—cf. § 4 of Chapter III), $X$ must be truth-functionally deducible from $M$.

(b) Conversely suppose $X$ is truth-functionally deducible from $M$, and that $X$ has no parameters. Then $X$ is truth-functionally implied by some finite subset $M_0$ of $M$, hence the set $M_0 \cup \{\sim X\}$ is not truth-functionally satisfiable. A fortiori, $M_0 \cup \{\sim X\}$ is not first order satisfiable (why?). Then $\{\sim X\}$ fails to be first-order satisfiable (because if it were, then $M_0 \cup \{\sim X\}$ would also be, by property $M_2$), which means that $X$ is valid. This concludes the proof.

We now define an *infinite* set to be regular if every finite subset is regular. A regular set $M$ will be called *complete* if for every $\gamma$, the sentence $\gamma \supset \gamma(a)$ is in $M$ for every parameter $a$, and for every $S$, there is at least one parameter $a$ such that $S \supset S(a)$ is in $M$.

**Theorem 2.** *There exists a complete regular set $M$, and every such set is a magic set.*

Before proving Theorem 2, we wish to explicitly note:

**Lemma 1.** *Let $v$ be a Boolean valuation with the following 2 properties (for every $\gamma, \delta$):*

(1) If $\gamma$ is true under $v$, then for every parameter $a$, $\gamma(a)$ is true under $v$.

(2) If $\delta$ is true under $v$, then for at least one parameter $a$, $\delta(a)$ is true under $v$.

Then $v$ is a first-order valuation.

**Proof.** We shall use the conventional notation "$\varphi(x)$" for any formula containing no free variables other than (possibly) $x$, and for any parameter $a$, "$\varphi(a)$" denotes $[\varphi(x)]_a^x$.

(*a*) Suppose $(\forall x)\varphi(x)$ is true under $v$. Then by (1), for every parameter $a$, $\varphi(a)$ is true under $v$. We must show that conversely, if for every $a$, $\varphi(a)$ is true under $v$, then $(\forall x)\varphi(x)$ is true under $v$. This is equivalent to saying that if $(\forall x)\varphi(x)$ is false (under $v$) then for at least one parameter $a$, $\varphi(a)$ is false. So suppose that $(\forall x)\varphi(x)$ is false. Then the formula $\sim(\forall x)\varphi(x)$ must be true (since $v$ is a Boolean valuation). Then by (2), there is a parameter $a$ such that the formula $\sim\varphi(a)$ is true. Then $\varphi(a)$ must be false.

(*b*) If $(\exists x)\varphi(x)$ is true (under $v$) then for at least one $a$, $\varphi(a)$ is true (by (2)). If $(\exists x)\varphi(x)$ is false, then $\sim(\exists x)\varphi(x)$ is true, hence (by (1)) $\sim\varphi(a)$ is true for *every* parameter $a$, hence $\varphi(a)$ is false for every $a$. Thus if for some $a$, $\varphi(a)$ is true, then $(\exists x)\varphi(x)$ is true. This concludes the proof.

For future reference, we wish to point out that Lemma 1 can be formulated in the following equivalent form: Call a set $S$ a *Boolean* truth set if it is a truth set in the sense of propositional logic (i. e. its characteristic function is a *Boolean* valuation). We might remark that condition $M_1$ of the definition of a magic set is equivalent to the condition that every Boolean truth set which includes $M$ is a First-Order truth set. Lemma 1 is thus equivalent to the following:

**Lemma 1'.** *Let $S$ be a Boolean truth set with the following 2 properties (for every $\gamma, \delta$):*

(1) *If $\gamma \in S$, then for every parameter $a$, $\gamma(a) \in S$.*

(2) *If $\delta \in S$, then for at least one parameter $a$, $\delta(a) \in S$.*

*Then $S$ is a first-order truth set.*

Now we return to the proof of Theorem 2. We use the following method of *pre-assigned parameters*. We enumerate all $\delta$-sentences in some denumerable sequence $\delta_1, \delta_2, ..., \delta_n, ...$ and we also consider all our *parameters* given in some fixed enumeration $b_1, b_2, ..., b_n, ...$ Now we define $a_1$ to be the *first* parameter (of the sequence $\{b_i\}$) which does not occur in $\delta_1$, $a_2$ the *next* parameter which does not occur in $\delta_1$ nor $\delta_2$, $a_3$ the next parameter which does not occur in $\delta_1, \delta_2, \delta_3, ..., a_{n+1}$ the first parameter after $a_1, ..., a_n$ which does not occur in $\delta_1, \delta_2, ..., \delta_n, \delta_{n+1}$, etc. Let $R_1$ be the set $\delta_1 \supset \delta_1(a_1),\ \delta_2 \supset \delta_2(a_2), ..., \delta_n \supset \delta_n(a_n), ...$ Clearly, $R_1$ is regular. Let $R_2$ be the set of *all* sentences $\gamma \supset \gamma(a)$ (for every $\gamma$ and every $a$). Then $R_1 \cup R_2$ is obviously completely regular, and this will be our desired set $M$.

To prove $(M_1)$ (of our definition of a magic set), suppose $v$ is a Boolean valuation satisfying $M$. Suppose $\gamma$ is true under $v$. Now for any $a$, $\gamma \supset \gamma(a)$ is true under $v$ (because $\gamma \supset \gamma(a)$ is an element of $M$). Then $\gamma(a)$ must be true under $v$ (because $\gamma$, $\gamma \supset \gamma(a)$ both are true under $v$ and $v$ is a Boolean valuation). Thus hypothesis (1) of Lemma 1 is satisfied. To prove hypothesis (2), if $\delta$ is true under $v$, then for at least one $a$, $\delta \supset \delta(a)$ is true (since it belongs to $M$), hence $\delta(a)$ is true. Thus hypothesis (2) of Lemma 1 holds, so $v$ is indeed a first-order valuation.

As for condition $(M_2)$, any finite subset $R$ of $M$ is regular, hence if $S$ is a finite (or for that matter even infinite) pure set which is satisfiable, then $R \cup S$ is again satisfiable by Theorem 1, Chapter VII. This concludes the proof of Theorem 2.

## § 2. Applications of Magic Sets

The mere existence of a magic set (even one which is not regular) immediately yields the First Order Compactness Theorem and the Skolem-Löwenheim Theorem as consequences of the Compactness Theorem for propositional logic by the following beautiful argument. We first note the trivial:

**Lemma 2.** *Any superset $S$ of $M$ (where $M$ is any magic set) which is truth-functionally satisfiable is also first-order satisfiable in a denumerable domain.*

**Proof.** By hypothesis there is a *Boolean* valuation (of $E^V$) which satisfies $S$. Then of course, $v$ also satisfies $M$. Then $v$ must be a first-order valuation (since $M$ is a magic set). So $S$ is satisfied by the first-order valuation $v$ (of the denumerable set $E^V$). This concludes the proof.

Now for the remarkable proof of the Compactness and Skolem-Löwenheim theorems.

Suppose that $S$ is a set of pure sentences such that every finite subset of $S$ is (first-order) satisfiable. Let $M$ be any magic set. We first assert that every finite subset $K$ of $M \cup S$ is satisfiable. For $K$ is the union of a finite subset $M_0$ of $M$ with a finite subset $S_0$ of $S$. By hypothesis, $S_0$ is satisfiable. Therefore, $M_0 \cup S_0$ is satisfiable (by property $M_2$ of magic sets). Since $K$ is first-order satisfiable, then *a fortiori* $K$ is *truth-functionally* satisfiable (since any First Order valuation satisfying $K$ is also a *Boolean* valuation satisfying $K$). Thus we have shown that every finite subset of $M \cup S$ is *truth-functionally* satisfiable. Then by the Compactness Theorem for *propositional* logic, the entire set $M \cup S$ is *truth-functionally* satisfiable. Then by the above lemma, the set $M \cup S$—and hence also the subset $S$—is first-order satisfiable in a denumerable domain. This concludes the proof.

Now we point out that the existence of a *regular* magic set immediately yields an alternative proof of the *weak* form of the Fundamental Theorem. For suppose $X$ is valid; let $M$ be a regular set. Then by Theorem 1, $M$ includes a finite subset $R$ which *truth-functionally* implies $X$, and of course, $R$ is a regular finite set. Thus every valid $X$ is truth-functionally implied by a finite regular set.

We know that even the weak form of the Fundamental Theorem is enough to yield the completeness of the axiom system $Q_1$. Thus the completeness of $Q_1$ can be alternatively proved by the use of magic sets (rather than by our earlier analytic methods).

Next we wish to show how we can use complete regular sets to give an alternative (and simple) proof of the *strong* form of the Fundamental Theorem. Let us first note that property $M_1$ of the definition of "magic sets" can be alternatively stated thus: For any Boolean valuation $v$ satisfying $M$, the set of all sentences true under $v$ is a *first-order truth set*. What we now need is the following closely related fact.

**Lemma 3.** *Let $M$ be a complete regular set, and for any sentence $X$, let $M_X$ be the set of all elements $Q \supset Q(a)$ of $M$ such that $Q$ is a weak subformula of $X$. Let $v$ be any Boolean valuation which satisfies $M_X$ and let $T_X$ be the set of all weak subformulas of $X$ which are true under $v$. Then $T_X$ is a Hintikka set.*

**Proof.** We leave to the reader the trivial verification that $T_X$ has properties $H_0, H_1, H_2$ of Hintikka sets. As for $H_3$, suppose $\gamma \in T_X$. Then $\gamma$ is both true (under $v$) and is a weak subformula of $X$. By the latter fact, $\gamma \supset \gamma(a)$ belongs to $M_X$ (for *every* $a$), hence is true under $v$, so $\gamma(a)$ is true under $v$. Also $\gamma(a)$ must be a weak subformula of $X$ (since $\gamma$ is), so $\gamma(a) \in T_X$. Proof of $H_4$ is analogous, and is left to the reader.

Now, to prove the *strong* form of the Fundamental Theorem using complete regular sets, we reason as follows. Let $M$ be a complete regular set and let $X$ be any pure valid sentence. Define $M_X$ as in Lemma 3. Let

$v$ be any Boolean valuation satisfying $M_X$. We wish to show that $X$ is true under $v$. Let $T_X$ be the set of all weak subformulas of $X$ which are true under $v$; we are to show that $X$ is a member of $T_X$. Since $X$, $\sim X$ are both weak subformulas of $X$ and since exactly one of $X$, $\sim X$ is true under $v$, then exactly one of $X$, $\sim X$ lies in $T_X$. Now $T_X$ is a Hintikka set (by Lemma 3), hence every element of $T_X$ is first-order satisfiable (indeed $T_X$ is simultaneously satisfiable by Hintikka's lemma). But $\sim X$ is not first-order satisfiable (since $X$ is valid), hence $\sim X$ cannot lie in $T_X$. Thus it must be $X$ that lies in $T_X$, so $X$ is true under $v$.

We have thus shown that for any valid pure sentence $X$, $X$ is truth-functionally implied by $M_X$. Then by the Compactness Theorem for propositional logic, $X$ must be truth-functionally implied by some *finite* subset $R$ of $M_X$. It is immediate from the definition of $M_X$ that $R$ is not only regular, but that $R$ is an associate of $\{\sim X\}$. This concludes the proof.

*Exercise.* In our construction of a complete regular set $M$, the set of parameters $a_1, a_2, \dots, a_n, \dots$ may fail to exhaust all members of $\{b_i\}$. Can you modify the construction so that *every* parameter appears as one of the $a_i$?

Chapter X

# Analytic versus Synthetic Consistency Properties

We have remarked earlier that magic sets emerged from the completeness proofs of Henkin and Hasenjaeger. In this chapter we wish to discuss the Henkin and Hasenjaeger completeness proofs and their relationship to the completeness proofs of earlier chapters. We conclude this chapter with a new proof of the Unifying Principle—which circumvents the necessity of appealing to *systematic* tableaux—and we discuss the essential differences and similarities of what are basically two types of completeness proofs; the one along the lines of Lindenbaum-Henkin, the other along the lines of Gödel-Herbrand-Gentzen.

## § 1. Synthetic Consistency Properties

We have earlier spoken much about *analytic* consistency properties $\Gamma$. Now let $\Delta$ be a property of sets (of sentences) which again is of finite character. And again we shall call a set $S$ $\Delta$-consistent ($\Delta$-inconsistent) if $S$ does (does not) have the property $\Delta$, and we shall call a sentence $X$ $\Delta$-provable if the set $\{\overline{X}\}$ is $\Delta$-inconsistent. Now we define $\Delta$ to be a *synthetic* consistency property if the following conditions hold (for every $S, \gamma, \delta$):

$B_0$: If $S$ is *truth-functionally* inconsistent (i.e. if some finite subset of $S$ is truth-functionally unsatisfiable), then $S$ is $\Delta$-inconsistent.

$B_3$: If $\{S,\gamma\}$ is $\Delta$-consistent, so is $\{S,\gamma,\gamma(a)\}$.

$B_4$: If $\{S,\delta\}$ is $\Delta$-consistent and if $a$ does not occur in $\{S,\delta\}$, then $\{S,\delta,\delta(a)\}$ is $\Delta$-consistent.

$B_5$: If $\{S,X)$ is $\Delta$-inconsistent and if $\{S,\overline{X}\}$ is $\Delta$-inconsistent, then $S$ is $\Delta$-inconsistent. Stated otherwise, if $S$ is $\Delta$-consistent, then for every sentence $X$, at least one of the sets $\{S,X\}$ or $\{S,\overline{X}\}$ is $\Delta$-consistent.

We have purposely skipped the subscripts 1, 2 in our numbering of conditions $B_0 - B_5$ to allow us to compare them more easily with the conditions $A_0 - A_4$ defining *analytic* consistency properties.

Condition $B_0$ is an obvious strengthening of $A_0$. Conditions $A_1$, $A_2$ have no counterpart in the definition of synthetic consistency properties. Conditions $B_3$, $B_4$ (for the quantifiers) are the same as $A_3$, $A_4$ respectively. And condition $B_5$ is totally absent from the definition of *analytic* consistency. We shall refer to this very important condition $B_5$ as the *cut* conditition.

Conditions $B_0$, $B_5$ at once imply the following condition which we will call $B_6$: If $S$ is $\Delta$-consistent and if some finite subset of $S$ truth-functionally implies $X$, then $\{S, X\}$ is $\Delta$-consistent.

**Proof.** If $\{S,X\}$ were not $\Delta$-consistent, then $\{S,\overline{X}\}$ would be $\Delta$-consistent (by $B_5$), but this is contrary to $B_0$ (since by hypothesis some finite subset of $\{S,\overline{X}\}$ is not truth-functionally satisfiable).

Let us now note

**Theorem 1.** *Every synthetic consistency property is also an analytic consistency property.*

**Proof.** It remains to show that conditions $A_1$, $A_2$ hold (reading "$\Delta$" for "$\Gamma$").

(1): Suppose $\{S,\alpha\}$ is $\Delta$-consistent. Since $\alpha$ truth-functionally implies $\alpha_1$, then $\{S,\alpha,\alpha_1\}$ is $\Delta$-consistent (by $B_6$). Again since $\alpha$ truth-functionally implies $\alpha_2$, then $\{S,\alpha,\alpha_1,\alpha_2\}$ is $\Delta$-consistent (by $B_6$).

(2): Suppose $\{S,\beta\}$ is $\Delta$-consistent. Then either $\{S,\beta,\beta_1\}$ or $\{S,\beta,\overline{\beta}_1\}$ is $\Delta$-consistent (by $B_5$). If the former, then certainly at least one of $\{S,\beta,\beta_1\}$ or $\{S,\beta,\beta_2\}$ is $\Delta$-consistent and we have won our point. So suppose the latter—i.e. that $\{S,\beta,\overline{\beta}_1\}$ is $\Delta$-consistent. We then wish to show that $\{S,\beta,\beta_2\}$ is $\Delta$-consistent. Well either, $\{S,\beta,\overline{\beta}_1,\beta_2\}$ or $\{S,\beta,\overline{\beta}_1,\overline{\beta}_2\}$ is $\Delta$-consistent (why?), but the latter is impossible (since $\{\beta,\overline{\beta}_1,\overline{\beta}_2\}$ is not truth-functionally satisfiable). Therefore $\{S,\beta,\overline{\beta}_1,\beta_2\}$ is $\Delta$-consistent, hence so is its subset $\{S,\beta,\beta_2\}$ (because $\Delta$-consistency is of finite character). This concludes the proof.

We have earlier asked the reader to verify that consistency in the axiom system $Q_1$ is an analytic consistency property. Actually it is just about as easy to verify that consistency in $Q_1$ is even a *synthetic* consistency property.

Now Henkin's completeness proof is for systems like $Q_1$, in which consistency within the system is immediately a *synthetic* consistency property. We henceforth let "$\Delta$" stand for any synthetic consistency property. Suppose $S$ is a pure set which is $\Delta$-consistent. Henkin's idea is to show that $S$ can be extended to a first-order truth set. Of course, we can speak of *maximal* $\Delta$-consistency, but in general it is *not* the case that a maximal $\Delta$-consistent set is a first-order truth set! There is another condition needed, which we now discuss.

We shall call a set $S$ *E-complete* (existentially complete) if for every $\delta \in S$, there is at least one parameter $a$ such that $\delta(a) \in S$. Now we assert

**Theorem 2.** *If $M$ is both maximally $\Delta$-consistent and E-complete, then $M$ is a first-order truth set.*

**Proof.** Suppose $M$ is maximally $\Delta$-consistent and E-complete. That $M$ is a *Boolean* truth set can be proved exactly as in Lemma 1 (preceding Lindenbaum's Theorem)—Chapter III (indeed conditions $B_0$, $B_5$ are exactly $L_0$, $L_1$ discussed immediately following the statement of this lemma). Next we show that the hypotheses of Lemma 1' of Chapter IX hold for $M$. Well, suppose $\gamma \in M$. Then $\{M, \gamma\} = M$, so $\{M, \gamma\}$ is $\Delta$-consistent. Then for every parameter $a$, $\{M, \gamma, \gamma(a)\}$ is $\Delta$-consistent (by $B_3$). But, $\{M, \gamma, \gamma(a)\} = \{M, \gamma(a)\}$ is $\Delta$-consistent. Then $\gamma(a) \in M$ (since $M$ is *maximally* $\Delta$-consistent). Thus $\gamma \in M$ implies that for every $a, \gamma(a) \in M$. Thus $M$ has the first of the two properties required by Lemma 1' (Chapter IX). The second property is simply E-completeness. Thus by Lemma 1', $M$ is a first-order truth set.

**Remark.** The reader may wonder why E-completeness is not a consequence of maximal $\Delta$-consistency. Well, suppose $M$ is maximally $\Delta$-consistent and that $\delta \in M$. Now, if there is any parameter $a$ which occurs in no element of $M$, then indeed by $B_4$ and the maximal $\Delta$-consistency of $M$, $\delta(a)$ would have to be an element of $M$. But it may be that every parameter occurs in some element of $M$, in which case we have no assurance that for some $a$, $\{M, \delta(a)\}$ is again $\Delta$-consistent.

Now, Henkin's idea is to show how a $\Delta$-consistent pure set $S$ can be extended to a set which is simultaneously maximally $\Delta$-consistent and E-complete. He does this by the following ingenious method.

We partition our denumerable set of parameters into denumerably many sets $A_1, A_2, \ldots, A_n, \ldots$, each of which is denumerable—or rather we look at the denumerable *sequence* $A_1, A_2, \ldots, A_n, \ldots$ (without repetitions) where each $A_i$ is itself a denumerable *sequence* $a_1^i, a_2^i, \ldots, a_n^i, \ldots$ of parameters (without repetitions). We let $E_0$ be the set of all *pure* sentences, and for each $n > 0$, let $E_n$ be the set of all sentences using parameters just from $A_1 \cup \cdots \cup A_n$. We let $E_\omega$ be the set of *all* sentences, and we note that $E_\omega = E_0 \cup E_1 \cdots \cup E_n \cup \cdots$.

Call a set $S$ $E$-complete *relative to a subset* $S'$ if, for all $\delta \in S'$, there is a parameter $a$ such that $\delta(a) \in S$.

**Lemma.** *Any $\Delta$-consistent subset $S$ of $E_n$ can be extended to a subset of $E_{n+1}$ which is again $\Delta$-consistent and also $E$-complete relative to $S$.*

**Proof.** Arrange all $\delta$-type elements of $S$ in a denumerable sequence $\delta_1, \delta_2, \ldots, \delta_n, \ldots$ (or a finite sequence, if there are only finitely many such elements). Now all parameters of $A_{n+1}$ are new to $S$, and $A_{n+1}$ is a sequence $b_1, b_2, \ldots, b_n, \ldots$. Also each $b_{i+1}$ is new to $\{S, b_1, \ldots, b_i\}$. By property $B_4$, the set $\{S, \delta_1(b_1)\}$ is $\Delta$-consistent. Then again by $B_4$, the set $\{S, \delta_1(b_1), \delta_2(b_2)\}$ is $\Delta$-consistent. By an obvious induction argument, for each $n > 0$, the set $\{S, \delta_1(b_1), \ldots, \delta_n(b_n)\}$ is $\Delta$-consistent. This implies that the set $\{S, \delta_1(b_1), \ldots, \delta_n(b_n), \ldots\}$ is $\Delta$-consistent (because $\Delta$ is a property of finite character). And this set $\{S, \delta_1(b_1), \ldots, \delta_n(b_n), \ldots\}$ is obviously $E$-complete relative to $S$ (and also a subset of $E_{n+1}$). This concludes the proof.

Now for Henkin's construction. Let $S$ be a *pure* set which is $\Delta$-consistent. By the above lemma, we can extend $S$ to a subset $S_1$ of $E_1$ which is $\Delta$-consistent and $E$-complete relative to $S$. Of course, $S_1$ may fail to be a *maximally* $\Delta$-consistent subset of $E_1$, but by Lindenbaum's construction (or Tukey's lemma—cf. Chapter III), $S_1$ can be extended to a maximally $\Delta$-consistent subset $S_1^*$ of $E_1$. However $S_1^*$ may fail to be $E$-complete relative to $S$ (because in extending $S_1$ to $S_1^*$ we might have added some $\delta$ without adding any $\delta(a)$!). But by the above lemma, we can extend $S_1^*$ to a $\Delta$-consistent subset $S_2$ of $E_2$ which is $E$-complete relative to $S_1^*$. However, $S_2$, though $\Delta$-consistent, may fail to be a maximally $\Delta$-consistent subset of $E_2$, so we extend it to a maximally $\Delta$-consistent subset $S_2^*$ of $E_2$. We thus alternate between Lindenbaum's construction and the construction of the above lemma, and generate a denumerable ascending sequence $S, S_1, S_1^*, S_2, S_2^*, \ldots, S_i, S_i^*, \ldots$ (ascending in the sense that $S \subseteq S_1 \subseteq S_1^* \subseteq \cdots \subseteq S_i \subseteq S_i^* \subseteq \cdots$) such that for each $i$, $S_{i+1}$ is $E$-complete relative to $S_i^*$ and $S_i^*$ is a maximally $\Delta$-consistent subset of $E_i$. Then Henkin takes the *union* $S \cup S_1 \cup S_1^* \cup \cdots \cup S_i \cup S_i^* \cup \cdots$. We leave it to the reader to verify that this union is both $E$-complete and maximally $\Delta$-consistent (i.e. is a maximally $\Delta$-consistent subset of $E_\omega$). Then by Theorem 2, this union is a first-order truth set, so the subset $S$ is first-order satisfiable (indeed in a denumerable domain). This concludes Henkin's proof.

**The Henkin-Hasenjaeger Proof.** A simplification of Henkin's proof —which avoids alternating between Lindenbaum's construction and $E$-completion—was discovered independently by Hasenjaeger, Henkin (and apparently also by Beth, and probably several others).

The key fact needed is the following

**Lemma.** *Suppose $S$ is $\Delta$-consistent. Then*
(1) $\{S, \gamma \supset \gamma(a)\}$ *is $\Delta$-consistent.*
(2) $\{S, \delta \supset \delta(a)\}$ *is $\Delta$-consistent, providing $a$ is new to $\{S, \delta\}$.*

**Proof.** Suppose $S$ is $\Delta$-consistent. Now consider any sentence $Q \supset Q(a)$, where $Q$ is either some $\gamma$ or some $\delta$ such that $a$ is new to $\{S, Q\}$. We wish to show that $\{S, Q \supset Q(a)\}$ is $\Delta$-consistent. Since $S$ is $\Delta$-consistent, then either $\{S, Q\}$ or $\{S, \sim Q\}$ is $\Delta$-consistent (by $B_5$). If $\{S, Q\}$ is $\Delta$-consistent, so is $\{S, Q, Q(a)\}$ (by $B_3$, if $Q$ is some $\gamma$, or by $B_4$ if $Q$ is some $\delta$—since in this case we are assuming that $a$ is new to $\{S, Q\}$), and hence the subset $\{S, Q(a)\}$ is $\Delta$-consistent. Thus either $\{S, \sim Q\}$ or $\{S, Q(a)\}$ is $\Delta$-consistent. If the former, then $\{S, Q \supset Q(a)\}$ is $\Delta$-consistent (by property $B_6$, since $\sim Q$ truth-functionally implies $Q \supset Q(a)$). If the latter, then again $\{S, Q \supset Q(a)\}$ is $\Delta$-consistent (by $B_6$, since $Q(a)$ truth-functionally implies $Q \supset Q(a)$). Thus in either case, $\{S, Q \supset Q(a)\}$ is $\Delta$-consistent.

**Remark.** The proof of the above lemma is essentially the same thing as the proof that everything provable in the system $Q_2$ is provable in the system $Q_1''$—indeed one can obtain both results from a common construction (how?).

Now we consider the Henkin-Hasenjaeger proof. Suppose $S$ is a pure set which is $\Delta$-consistent. Arrange *all* $\delta$-sentences (not just those in $S$!) in some denumerable sequence $\delta_1, \delta_2, ..., \delta_n, ...$. By the above lemma, we can take any parameter $a_1$ not in $\delta_1$, and adjoin to $S$ the sentence $\delta_1 \supset \delta_1(a_1)$, and the resulting set $S_1$ is again $\Delta$-consistent. Then take a parameter $a_2$ not in $\{S_1, \delta_2\}$ and $S_1 \cup \{\delta_2 \supset \delta_2(a_2)\}$ is again $\Delta$-consistent. So we inductively define the sequence $S_1, ..., S_n, ...$ by the conditions: $S_0 = S$; $S_{i+1} = S_i \cup \{\delta_{i+1} \supset \delta_{i+1}(a_{i+1})\}$ where $a_i$ is some parameter (say the first in some fixed enumeration) which does not occur in $(S_i, \delta_{i+1})$. Then each $S_i$ is $\Delta$-consistent, so their union $S^*$ is $\Delta$-consistent. Now, $S^*$ has the property that any superset $S^0$ of $S^*$ which is closed under truth-functional implication must be $E$-complete (because if $\delta \in S^0$, then for some parameter $a$, $\delta \supset \delta(a) \in S^0$, hence by the closure under truth-functional implication, $\delta(a) \in S^0$). Now take for $S^0$ a *maximally* consistent extension of $S^*$. Then $S^0$ must be closed under truth-functional implication (why?), hence $S^0$ is $E$-complete. Hence $S^0$ is a first-order truth set.

**Discussion.** In effect, the Henkin-Hasenjaeger construction manufactures "half" of a magic set in the course of the proof, and adjoins it to $S$ and obtains $S^*$. At this point, one could modify the remainder of the proof as follows: Instead of extending $S^*$ to a maximal consistent set, we could alternatively add to $S^*$ all sentences of the form $\gamma \supset \gamma(a)$; the resulting set $S^{**}$ is again consistent by (1) of the Lemma. Then $S^{**}$ is truth-functionally satisfiable, so we take a Boolean valuation $v$ satisfying $S^{**}$ and then the set $S^{***}$ of all sentences true under $v$ is a *first-order truth set* (because $S^{**}$ is a superset of a magic set). Essentially, therefore, the Henkin-Hasenjaeger proof boils down to the facts: (*i*) for any $\Delta$-consistent pure set $S$ and any magic set $M$, the set $S \cup M$ is again $\Delta$-consistent (and this follows from the above Lemma by induction); (*ii*) any truth-functionally satisfiable superset of a magic set is first-order satisfiable.

## § 2. A More Direct Construction

Professor Henkin has pointed out to the author another completeness proof which is strikingly direct. And this proof can be modified to yield another proof of our Unifying Principle (and hence also of the Completeness Theorem for tableaux) which completely avoids the necessity for considering "systematic" tableaux.

The problem, as before, is how to extend a $\Delta$-consistent pure set $S$ to a first-order truth set. Let us recall Lindenbaum's construction: We first enumerate all sentences in a sequence $X_1, \ldots, X_n, \ldots$, and at the $n$-th stage of our construction, we take the set $S_n$ already at hand, and we adjoin $X_{n+1}$ provided that this does not destroy $\Delta$-consistency, otherwise we leave $X_{n+1}$ out. Now Henkin suggests the following modification: If we do adjoin $X_{n+1}$, *and if* $X_{n+1}$ *happens to be some* $\delta$, then we also throw in $\delta(a)$ for some $a$ new to $S_n$. Then the union $S \cup S_1 \cup \cdots \cup S_n \cup \cdots$ is without further ado both maximally $\Delta$-consistent and $E$-complete (verify!).

Thus in one construction, we have simultaneously achieved maximal consistency and $E$-completeness. It is hard to imagine a more direct completeness proof!

Now for the modification of the proof for *analytic* consistency properties $\Gamma$. Of course we can perform the same construction, and the resulting set $S \cup S_1 \cup \cdots \cup S_n \cup \cdots$ will indeed be both maximally $\Gamma$-consistent and $E$-complete. But such a set is not necessarily a first-order truth set! However, we have the following:

**Theorem.** *Let* $S^0$ *be the set of all subformulas of a pure set* $S$, *and let* $M$ *be a maximally* $\Gamma$-*consistent subset of* $S^0$ *which is also* $E$-*complete. Then* $M$ *is a Hintikka set.*

**Proof.** The proof that $M$ satisfies the truth-functional conditions $H_0, H_1, H_2$ of the definition of "Hintikka set" is the same as that of the theorem of § 3, Chapter III. As for $H_3$, suppose $\gamma \in M$. Then for every $a$, the set $\{M, \gamma(a)\}$ is $\Gamma$-consistent. But $\gamma(a) \in S^0$ (since $\gamma \in S^0$), so $\gamma(a) \in M$ (by the fact that $M$ is maximally $\Gamma$-consistent relative to $S^0$). This proves $H_3$. Condition $H_4$ is immediate from the hypothesis of $E$-completeness. This concludes the proof.

**Remark.** Strictly speaking, the above theorem is correct for the case when we are working with *signed* formulas. If we work with *unsigned* formulas, then we must introduce the notion of "descendant" in First-Order Logic, as we did for propositional logic (cf. § 3, Chapter III). We do this by simply adding to the definition of *direct descendant* (§ 3, Chapter III) "or $X$ is some $\gamma$ and $Y$ is $\gamma(a)$ for some $a$, or $X$ is some $\delta$ and $Y$ is $\delta(a)$ for some $a$". Then the above theorem is correct, reading "descendant" for "subformula".

Now the modification of Henkin's idea should be obvious. Suppose $S$ is a $\Gamma$-consistent pure set. Let $S^0$ be the set of all subformulas (or descendents, if we are working with unsigned formulas) of elements of $S$, and enumerate $S^0$ in a denumerable sequence $X_1, \ldots, X_n, \ldots$. Then carry out the same construction, and the resulting set $S \cup S_1 \cup \cdots \cup S_n \cup \cdots$ will be maximally $\Gamma$-consistent relative to $S^0$ and $E$-complete. By the above theorem it is a Hintikka set, hence by Hintikka's lemma it is satisfiable, hence the subset $S$ is satisfiable (and indeed in a denumerable domain).

We now see clearly that there are basically two types of completeness proofs. The first (along the Lindenbaum-Henkin lines) extends the "consistent" set directly to a truth set, and for this we need the full force of *synthetic* consistency properties (in particular the cut condition plays an essential role). And this method is directly applicable to the usual Hilbert-type formalization of First-Order Logic (such as the system $Q_1$). The second type of completeness proof (which we term "analytic", and is along the lines of Gödel, Herbrand, Gentzen) extends the "consistent" set, not directly to a truth set, but to a Hintikka set, which then can be further extended to a truth set. And for this method to work, we do not need the cut condition for *synthetic* consistency properties, but only the conditions of *analytic* consistency properties.

# Part III

# Further Topics in First-Order Logic

Chapter XI

# Gentzen Systems

## § 1. Gentzen Systems for Propositional Logic

We have written this section so that it can be read directly following Chapter II.

*Block Tableaux.* Preparatory to the main subject of this chapter, it will be convenient to consider a variant of the tableau method.

Our method of analytic tableaux is a variant of the tableau method of Beth. The *block* tableaux, to which we now turn, are substantially the tableaux of Hintikka [1]. In these tableaux, the "points" of the tree are finite sets of formulas rather than single formulas. And what we can do at any stage of the construction is dependent solely on the *end* points of the tree.

By a *block tableau* for a finite set $K$, we mean a tree constructed by placing the set $K$ at the origin, and then continuing according to the following rules:

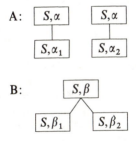

In words the rules above are:

A: To any end point $\{S,\alpha\}$ we may adjoin $\{S,\alpha_1\}$ as sole successor, or we may adjoin $\{S,\alpha_2\}$ as sole successor.

B: To any end point $\{S,\beta\}$ we may simultaneously adjoin $\{S,\beta_1\}$ as left successor and $\{S,\beta_2\}$ as right successor.

We call a block tableau *closed* if each end point contains some element and its conjugate, and *atomically closed* if each end point contains some atomic element and its conjugate.

Let us carefully note that in Rule $A$, we allow the possibility that $\alpha$ may be an element of $S$, and in Rule $B$, that $\beta$ may be an element of $S$.

Thus the following rules are but special cases of rules $A$ and $B$:

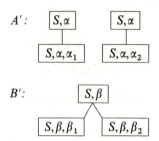

*Example of a Block Tableau.* The formula $p \supset r$ is truth-functionally implied by the set $p \supset q$, $q \supset r$—equivalently, the set $\{Tp \supset q,\ Tq \supset r,\ Fp \supset r\}$ is unsatisfiable. The following is a closed block tableau for this set:

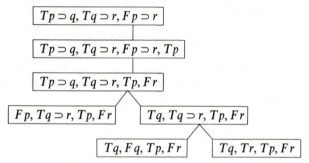

*Completeness.* Converting a closed *analytic* tableau for a set $S$ into a closed *block* tableau for $S$ is a very easy matter. Consider first the case that $S$ contains only one sentence $X$; let $\mathcal{T}$ be a closed analytic tableau for $\{X\}$. Then if we simply replace each point of $\mathcal{T}$ by the set consisting of the point itself together with all points above it on the branch, the resulting tree $B$ is closed block tableau for $\{X\}$.

*Example.* The following is a closed analytic tableau for $F(p \vee q) \supset (q \vee p)$:

(1)  $F(p \vee q) \supset (q \vee p)$

(2)  $T(p \vee q)$

(3)  $F(q \vee p)$

(4)  $Fq$

(5)  $Fp$

(6)  $Tp$      (7)  $Tq$

The following is its corresponding block tableau:

(1) $\boxed{F(p \vee q) \supset (q \vee p)}$

(2) $\boxed{F(p \vee q) \supset (q \vee p), T(p \vee q)}$

(3) $\boxed{F(p \vee q) \supset (q \vee p), Tp \vee q, Fq \vee p}$

(4) $\boxed{F(p \vee q) \supset (q \vee p), Tp \vee q, Fq \vee p, Fq}$

(5) $\boxed{F(p \vee q) \supset (q \vee p), Tp \vee q, Fq \vee p, Fq, Fp}$

(6) $\boxed{F(p \vee q) \supset (q \vee p), Tp \vee q, Fq \vee p, Fq, Fp, Tp}$

(7) $\boxed{F(p \vee q) \supset (q \vee p), Tp \vee q, Fq \vee p, Fq, Fp, Tq}$

We have just shown how to convert a closed analytic tableau for a single formula into a closed block tableau for that formula. Now let $S$ be a finite set $\{X_1, \ldots, X_n\}$ (where perhaps $n > 1$), and let $\mathcal{T}$ be a closed analytic tableau for $S$. To obtain a closed block tableau for $S$, we proceed as before, but we then delete the origin $\{X_1\}$, its successor $\{X_1, X_2\}, \ldots$, its successor $\{X_1, \ldots, X_{n-1}\}$.

*Example.* The following is a closed analytic tableau for the set $\{Tp \wedge q, Fp\}$:

(1) $Tp \wedge q$

(2) $Fp$

(3) $Tp$

Replacing each point of the tree by the set of all points which dominate it, we obtain the tree:

(1) $\boxed{Tp \wedge q}$

(2) $\boxed{Tp \wedge q, Fp}$

(3) $\boxed{Tp \wedge q, Fp, Tp}$

This is not a block tableau for $\{Tp \wedge q, Fp\}$, but if we delete point (1), we obtain the desired block tableau:

$\boxed{Tp \wedge q, Fp}$

$\boxed{Tp \wedge q, Fp, Tp}$

*Modified Block Tableaux.* The following modification of block tableaux will be useful.

Rule *A* of our rules for constructing block tableaux is really 2 rules (one for $\alpha_1$ and one for $\alpha_2$). Let us now replace this by the following single rule:

$$\boxed{S,\alpha}$$
$$|$$
$$\boxed{S,\alpha_1,\alpha_2}$$

Call a tree constructed using modified Rule A in place of the original Rule A a *modified* block tableau.

Any closed modified block tableau for *S* can be easily converted into a closed block tableau for *S* by replacing each part of the tree:

$$\boxed{R,\alpha}$$
$$|$$
$$\boxed{R,\alpha_1,\alpha_2}$$

by

$$\boxed{R,\alpha}$$
$$|$$
$$\boxed{R,\alpha,\alpha_1}$$
$$|$$
$$\boxed{R,\alpha_1,\alpha_2}$$

(In other words, 2 applications of the original Rule A can be used to accomplish the same thing as one application of the modified Rule A).

To convert a closed block tableau $\mathscr{B}$ for *S* into a closed modified block tableau for *S*, we replace each point $\{R,\alpha_1\}$ of $\mathscr{B}$ which was obtained from its predecessor $\{R,\alpha\}$ by Rule A by the point $\{R,\alpha_1,\alpha_2\}$ and also add $\alpha_2$ to each point of the tree below that point. The resulting $\mathscr{B}'$ is not quite a modified tableau, since certain points instead of being derived from their predecessors by Rule A or Rule B may simply be *repetitions* of their predecessors. But if we simply delete these repetitions from the tree, the resulting tree $\mathscr{B}''$ is then a modified block tableau for *S*. And of course, $\mathscr{B}''$ is closed (assuming $\mathscr{B}$ is closed).

*Gentzen Sequents.* By a *sequent* we shall mean an ordered pair $\langle U,V \rangle$ of finite sets of unsigned formulas. We shall use the more suggestive notation $U \rightarrow V$ for the sequent $\langle U,V \rangle$. Informally we read $U \rightarrow V$ as saying "if all elements of *U* are true, then at least one element of *V* is true". More precisely, we say that a sequent $U \rightarrow V$ is *true* under an interpretation (or Boolean valuation) *I* if either some element of *U* is false under *I* or some element of *V* is true under *I*—or what is the same thing, if all elements of *U* are true under *I*, then at least one element of

$V$ is true under $I$. Thus under any interpretation, the sequent $X_1, ..., X_n$ $\rightarrow Y_1, ..., Y_k$ (i.e. the sequent $\langle \{X_1, ..., X_n\}, \{Y_1, ..., Y_k\} \rangle$) has the same truth value as the formula $(...(X_1 \wedge X_2) \wedge \cdots \wedge X_n) \supset (...(Y_1 \vee Y_2) \vee \cdots \vee Y_k)$. We call $U \rightarrow V$ a *tautology* iff it is true under all interpretations; *satisfiable* iff it is true in at least one.

The elements of $U$ and the elements of $V$ are called the *terms* of the sequent $U \rightarrow V$. We allow the case that one or both sides of the arrow (i.e. one or both of the sets $U, V$) may be the empty set $\phi$. Following GENTZEN we use $\rightarrow Y_1, ..., Y_k$ to mean $\phi \rightarrow Y_1, ..., Y_k$, and $X_1, ..., X_n \rightarrow$ to mean $X_1, ..., X_n \rightarrow \phi$, and finally $\rightarrow$ to mean $\phi \rightarrow \phi$. We note that $\rightarrow V$ is true (under a given interpretation) iff at least one element of $V$ is true; $U \rightarrow$ is true iff at least one element of $U$ is false, and finally $\rightarrow$ is false in any interpretation (because the left side contains no member which is false—indeed it contains no members at all—and the right side contains no member which is true). Thus the sequent $\phi \rightarrow \phi$ is unsatisfiable.

By a tableau for a sequent $X_1, ..., X_n \rightarrow Y_1, ..., Y_k$ is meant a tableau for the set $\{TX_1, ..., TX_n, FY_1, ..., FY_k\}$. Note that the sequent $X_1, ..., X_n$ $\rightarrow Y_1, ..., Y_k$ is a tautology iff the set $\{TX_1, ..., TX_n, FY_1, ..., FY_k\}$ is unsatisfiable. Accordingly, a closed tableau for the set $\{TX_1, ..., TX_n, FY_1, ..., FY_k\}$ is also called a *tableau-proof* of the sequent $X_1, ..., X_n$ $\rightarrow Y_1, ..., Y_k$.

*The Axiom System $\mathscr{G}_0$.* Now we consider an *axiom system $\mathscr{G}_0$* in which the formal objects are sequents. This system is a somewhat modernized version of Gentzen's original (cf. the discussion at the end of §1). We have one axiom scheme, and eight inference rules—2 for each of the logical connectives; one for the introduction of the connective in the antecedent (left side of the arrow) and one for the introduction in the succedent (right side of the arrow). The postulates are as follows:

*Axioms.* $U, X \rightarrow V, X$

The rules are:

*Conjunction*   $C_1: \dfrac{U, X, Y \rightarrow V}{U, X \wedge Y \rightarrow V}$

$C_2: \dfrac{U \rightarrow V, X \quad U \rightarrow V, Y}{U \rightarrow V, X \wedge Y}$

*Disjunction*   $D_1: \dfrac{U \rightarrow V, X, Y}{U \rightarrow V, X \vee Y}$

$D_2: \dfrac{U, X \rightarrow V \quad U, Y \rightarrow V}{U, X \vee Y \rightarrow V}$

*Implication*         $I_1$:  $\dfrac{U, X \to V, Y}{U \to V, X \supset Y}$

                      $I_2$:  $\dfrac{U \to V, X \quad U, Y \to V}{U, X \supset Y \to V}$

*Negation*            $N_1$:  $\dfrac{U, X \to V}{U \to V, \sim X}$

                      $N_2$:  $\dfrac{U \to V, X}{U, \sim X \to V}$

It is obvious that the axioms of $\mathscr{G}_0$ are tautologies, and we leave it to the reader to verify that all the inference rules preserve tautologies—indeed in any application of any inference rule, the conclusion is truth-functionally implied by its premises. Therefore all theorems of $\mathscr{G}_0$ are tautologies. We will shortly prove the *completeness* of $\mathscr{G}_0$—i.e. that every sequent $U \to V$ which is a tautology is provable in $\mathscr{G}_0$.

*The System $\mathscr{G}_0$ in Uniform Notation.* Now we shall again exploit our unifying "$\alpha, \beta$" notation, which will enable us to collapse the eight inference rules of $\mathscr{G}_0$ to two.

Let $S$ be a set $\{TX_1, \ldots, TX_n, FY_1, \ldots, FY_k\}$ of signed formulas. By $|S|$ we shall mean the sequent $X_1, \ldots, X_n \to Y_1, \ldots, Y_k$. The correspondence between sets $S$ of signed formulas and their corresponding sequents $|S|$ is obviously $1-1$.

Now we can reformulate the postulates of $\mathscr{G}_0$ in uniform notation as follows:

*Axioms.*  $|S, TX, FX|$

*Inference Rules.*

                      $A$:  $\dfrac{|S, \alpha_1, \alpha_2|}{|S, \alpha|}$

                      $B$:  $\dfrac{|S, \beta_1| \, |S, \beta_2|}{|S, \beta|}$

To see that this really is the system $\mathscr{G}_0$, we let $U$ be the set of all $X$ such that $TX$ lies in $S$, and $V$ the set of all $Y$ such that $FY$ lies in $S$. We also let $X_1, \ldots, X_n$ be the members of $U$ and $Y_1, \ldots, Y_k$ the members of $V$ (thus $S = \{TX_1, \ldots, TX_n, FY_1, \ldots, FY_k\}$, and $|S| = X_1, \ldots, X_n, \to Y_1, \ldots, Y_k$). Then the axioms are all sequents of the form $|TX_1, \ldots, TX_n, FY_1, \ldots, FY_k,$ $TX, FX|$—i.e. all sequents $X_1, \ldots, X_n, X \to Y_1, \ldots, Y_k, X$—which is all sequents $U, X \to V, X$.

As for the inference rules, we must consider separately the various $\alpha$ and $\beta$ cases. Suppose, for example, that $\alpha = TX \wedge Y$. This case of Rule A is then "$|TX_1,...,TX_n,TX,TY,FY_1,...,FY_k|$ yields $|TX_1,...,TX_n$, $T(X \wedge Y), FY_1,...,FY_k|$", which is "$U,X,Y \rightarrow V$ yields $U,X \wedge Y \rightarrow V$", which is rule $C_1$. The reader can now easily verify that if we take $FX \vee Y$, $FX \supset Y$, $F \sim X$ for $\alpha$, we respectively get rules $D_1, I_1, N_1$. If we take $FX \wedge Y$, $TX \vee Y$, $TX \supset Y$, $T \sim X$ for $\beta$, we respectively get rules $C_2, D_2$, $I_2, N_2$. (Also if we take $T \sim X$ for $\alpha$, we get $N_2$; if we take $F \sim X$ for $\beta$, we get $N_1$.) Thus our uniform presentation of $\mathscr{G}_0$ is correct.

We might remark that in our uniform presentation of $\mathscr{G}_0$, we needed only one metalinguistic variable "$S$", whereas in the prior formulation we needed two metalinguistic variables "$U$", "$V$".

*Completeness of* $\mathscr{G}_0$. Suppose $X_1,...,X_n \rightarrow Y_1,...,Y_k$ is a tautology. Let $\mathscr{M}$ be a closed modified block tableau for the set $\{TX_1,...,TX_n$, $FY_1,...,FY_k\}$. To transfer $\mathscr{M}$ into a proof (in tree form) of $X_1,...,X_n \rightarrow Y_1,...,Y_k$ in the axiom system $\mathscr{G}_0$ all we need do is replace each point $S$ of $\mathscr{M}$ by the sequent $|S|$ (but the resulting proof tree is usually displayed upside down—i.e. with the origin at the bottom). Thus the system $\mathscr{G}_0$ is complete.

*Example.* The following is a closed modified block tableau for the sequent $p \supset q \rightarrow \sim q \supset \sim p$:

$$Tp \supset q, F \sim q \supset \sim p$$
$$Tp \supset q, T \sim q, F \sim p$$
$$Tp \supset q, Fq, F \sim p$$
$$Tp \supset q, Fq, Tp$$
$$Fp, Fq, Tp \qquad Tq, Fq, Tp$$

Its corresponding proof in $\mathscr{G}_0$ is as follows:

$$p \rightarrow q, p \qquad p, q \rightarrow q$$
$$p, p \supset q \rightarrow q$$
$$p \supset q \rightarrow q, \sim p$$
$$p \supset q, \sim q, \rightarrow \sim p$$
$$p \supset q \rightarrow \sim q \supset \sim p$$

**Discussion.** We have remarked that the system $\mathscr{G}_0$ is a modification of the original system of Gentzen—it is substantially that which appears in Lyndon [1]. In the original Gentzen formulation, he worked with sequents $U \rightarrow V$ in which $U, V$ instead of being finite *sets* of formulas are finite *sequences* of formulas. Instead of the axiom scheme $U, X \rightarrow V, X$, Gentzen used the simpler scheme $X \rightarrow X$. But also Gentzen added the following three so-called *structural rules:*

$$\text{Thinning:} \quad \frac{U \rightarrow V}{U \rightarrow V, X} \qquad\qquad \frac{U \rightarrow V}{U, X \rightarrow V}$$

$$\text{Contraction:} \quad \frac{U \rightarrow V, X, X}{U \rightarrow V, X} \qquad\qquad \frac{U, X, X \rightarrow V}{U, X \rightarrow V}$$

$$\text{Interchange:} \quad \frac{U \rightarrow V_1, X, Y, V_2}{U \rightarrow V_1, Y, X, V_2} \qquad \frac{U_1, X, Y, U_2 \rightarrow V}{U_1, Y, X, U_2 \rightarrow V}$$

The system—call it $\mathscr{G}_*$—is equivalent to $\mathscr{G}_0$ in the following sense: Let $\Theta$, $\Gamma$ be finite *sequences* of formulas; let $U$ be the *set* of terms of $\Theta$ and $V$ be the set of terms of $\Gamma$. Then $\Theta \rightarrow \Gamma$ is provable in $\mathscr{G}_*$ iff $U \rightarrow V$ is provable in $\mathscr{G}_0$.

Several modifications of $\mathscr{G}_0$ exist. For example, let $\mathscr{G}_1$ be the system obtained from $\mathscr{G}_0$ by replacing rules $C_1, D_1$ as follows: $C_1$ is replaced by the two rules:

$$\frac{U, X \rightarrow V}{U, X \wedge Y \rightarrow V} \qquad\qquad \frac{U, Y \rightarrow V}{U, X \wedge Y \rightarrow V}$$

$D_1$ is replaced by the two rules:

$$\frac{U \rightarrow V, X}{U \rightarrow V, X \vee Y} \qquad\qquad \frac{U \rightarrow V, Y}{U \rightarrow V, X \vee Y}$$

The system $\mathscr{G}_1$ is not uniform—i.e. it is not possible to present the inference rules of $\mathscr{G}_1$ in the uniform $\alpha, \beta$-notation. It can be made uniform by also replacing Rule $I_1$ by the two rules:

$$\frac{U, X \rightarrow V}{U \rightarrow V, X \supset Y} \qquad\qquad \frac{U \rightarrow V, Y}{U \rightarrow V, X \supset Y}$$

The resulting system—call it $\mathscr{G}_2$—is uniform—it can be presented uniformly by replacing Rule $A$ by the two rules:

$$\frac{|S, \alpha_1|}{|S, \alpha|} \qquad\qquad \frac{|S, \alpha_2|}{|S, \alpha|}$$

The system $\mathscr{G}_2$ bears roughly the same relationship to block tableaux as $\mathscr{G}_0$ does to modified block tableaux. Indeed, if in a closed block tableau we replace each point $S$ by the sequent $|S|$, the resulting tree is a proof in $\mathscr{G}_2$.

## § 2. Block Tableaux and Gentzen Systems for First-Order Logic

To extend the system of block tableaux (either modified or unmodified) to Quantification Theory, we add the rules:

$C$:

$$\boxed{S, \gamma}$$
$$|$$
$$\boxed{S, \gamma, \gamma(a)}$$

$D$:    $\boxed{S, \delta}$ ,     provided $a$ is new.

$$\boxed{S, \delta(a)}$$

The procedure for converting a closed analytic tableau for $S$ into a closed block tableau for $S$ is exactly the same as for propositional logic.

To extend the Gentzen system $\mathscr{G}_0$ to First-Order Logic, we add the following rules ($\varphi(x)$ is any sentence, $\varphi(a)$ is the result of substituting $a$ for all free occurrences of $x$ in $\varphi(x)$):

$$U_1: \quad \frac{U, \varphi(a) \to V}{U, (\forall x)\varphi(x) \to V} \qquad U_2: \quad \frac{U \to V, \varphi(a)}{U \to V, (\forall x)\varphi(x)}$$

(provided $a$ does not occur in the conclusion)

$$\exists_1: \quad \frac{U \to V, \varphi(a)}{U \to V, (\exists x)\varphi(x)} \qquad \exists_2: \quad \frac{U, \varphi(a) \to V}{U, (\exists x)\varphi(x) \to V}$$

(with same proviso)

In uniform notation, the quantificational rules are:

$$C: \quad \frac{|S, \gamma(a)|}{|S, \gamma|}$$

$$D: \quad \frac{|S, \delta(a)|}{|S, \delta|}, \text{ provided } a \text{ does not occur in any term of } \{S. \delta\}.$$

We shall use the letter "$\mathscr{G}$" for the above system. The completeness of $\mathscr{G}$ can be proved from the completeness of modified block tableaux in exactly the same manner as in propositional logic: If in a closed modified block tableau for $K$ we replace each point $S$ by the sequent $|S|$, we obtain a proof in $\mathscr{G}$ of the sequent $|K|$.

Of course, one can alternatively obtain the completeness of $\mathscr{G}$ directly as a consequence of our unifying principle: Define $S$ to be *consistent* if the sequent $|S|$ is not provable in $\mathscr{G}$. It is immediate from the postulates of $\mathscr{G}$ that we then have an analytic consistency property.

For subsequent applications, it is important to note that the system $\mathscr{G}$ remains complete if we restrict our axiom scheme, $U, X \rightarrow V, X$ to the case that $X$ is *atomic*. For we know that if $S$ is unsatisfiable, then there is an *atomically* closed modified block tableau for $S$. The corresponding proof of $|S|$ in $\mathscr{G}$ will then use only this weaker axiom scheme.

## Chapter XII

# Elimination Theorems

### § 1. Gentzen's Hauptsatz

Call $X$ *eliminable* if for every finite set $S$, if there exists a closed tableau for $\{S, X\}$ and a closed tableau for $\{S, \bar{X}\}$, then there exists a closed tableau for $S$. It is an immediate corollary of the Completeness theorem for tableaux that every $X$ is eliminable. For suppose there is a closed tableau for $\{S, X\}$ and a closed tableau for $\{S, \bar{X}\}$. Then both $\{S, X\}$ and $\{S, \bar{X}\}$ are unsatisfiable, hence $S$ is unsatisfiable (because in any interpretation at least one of $X$, $\bar{X}$ is true). Then by the Completeness theorem, there must be a closed tableau for $S$.

Gentzen's Hauptsatz (or rather its counterpart for tableaux) is that every $X$ is eliminable. The above proof—using the Completeness theorem—is of course non-constructive. It provides no direct method whereby given a closed tableau $\mathscr{J}_1$ for $\{S, X\}$ and a closed tableau $\mathscr{J}_2$ for $\{S, \bar{X}\}$ we can find a closed tableau $\mathscr{J}$ for $S$. Of course, given a closed tableau $\mathscr{J}_1$ for $\{S, X\}$ and a closed tableau $\mathscr{J}_2$ for $\{S, \bar{X}\}$ we could simply *ignore* $\mathscr{J}_1$ and $\mathscr{J}_2$ and set to work constructing a *systematic* tableau for $S$, and we know (by the Completeness theorem) that sooner or later our systematic tableau will close. But we have absolutely no idea of how long the systematic tableau for $S$ will run before closing. By contrast, Gentzen's proof of the Hauptsatz provides a wholly constructive upper bound for the complexity of $\mathscr{J}$, as a *function* $\varphi$ of the complexity of the given tableaux $\mathscr{J}_1$, $\mathscr{J}_2$.[1]

Gentzen, of course, proved the Hauptsatz not for tableaux (which were developed later), but for Gentzen systems. More specifically, consider the system $\mathscr{G}$ of the last chapter. Gentzen's form of the Hauptsatz is to the effect that if $U, X \rightarrow V$ and $U \rightarrow X, V$ are both provable in $\mathscr{G}$,

---

[1] For readers with some familiarity with recursive function theory, Gentzen's method provides a *primitive* recursive function $\varphi$ rather than a general recursive $\varphi$.

then $U \to V$ is provable in $\mathscr{G}$ (in fact more generally: if $U_1, X \to V_1$ and $U_2 \to X, V_2$ are both provable in $\mathscr{G}$, then $U_1, U_2 \to V_1, V_2$ is provable in $\mathscr{G}$). To be more accurate, Gentzen considered a system $\mathscr{G}^*$ like $\mathscr{G}$ together with the following rule:

(Cut Rule) $$\frac{U_1, X \to V_1 \quad U_2 \to X, V_2}{U_1, U_2 \to V_1, V_2}$$

A proof in $\mathscr{G}^*$ is called *cut-free* if it does not employ the cut rule. The precise form of Gentzen's formulation is that any sequent provable in $\mathscr{G}^*$ has a cut-free proof in $\mathscr{G}^*$ (i.e. a proof in $\mathscr{G}$).

The Hauptsatz for tableaux easily implies the Hauptsatz for Gentzen systems (by using constructive translation processes from proofs by tableaux to proofs in Gentzen systems, and vice versa). We find tableaux more pleasant to deal with than Gentzen systems, hence our proof of the Hauptsatz (which is basically that of Gentzen, but rather more simple in many respects) will be motivated largely by tableau considerations. Actually we shall prove the Hauptsatz in a more abstract form, which simultaneously yields the results both for tableaux and Gentzen's systems without necessitating translation devices.

## § 2. An Abstract Form of the Hauptsatz

All sets $S$ with which we now deal will be assumed *finite*. We shall say that $S$ is *closed* if $S$ contains some element and its conjugate. We shall say that $S$ *closes* if there exists a closed tableau for $S$. For any natural number $k$, we shall say that $S$ *closes with weight* $k$ if there is a closed tableau $\mathscr{T}$ for $S$ such that $k$ is the number of points of $\mathscr{T}$ *exclusive of the elements of* $S$. Thus, e.g., a set $S$ is closed iff $S$ closes with weight 0.

For any $k > 0$, we shall say that $S$ *closes via* $Y$ *with weight* $k$ if there is a closed tableau $\mathscr{T}$ for $S$ of weight $k$ such that $Y$ is the element of $S$ which was first used in the construction of $\mathscr{T}$. Obviously for any $k > 0$, if $S$ closes with weight $k$, then there must be an element $Y$ of $S$ such that $S$ closes via $Y$ with weight $k$ (but the $Y$ is not necessarily unique, since there might be two different closed tableaux $\mathscr{T}_1, \mathscr{T}_2$ for $S$, both of weight $k$, and $Y_1$ may be the element of $S$ first used in $\mathscr{T}_1$ and $Y_2$ may be the element of $S$ first used in $\mathscr{T}_2$).

Now we wish to consider a purely abstract 3-place relation $C(S, Y, k)$ between finite sets $S$, elements $Y$ and positive integers $k$, which we read "$S$ *closes via* $Y$ *with weight* $k$". Define "$S$ *closes with weight* $k$" to mean that either $k = 0$ and $S$ is closed, or $k > 0$ and there exists some non-atomic $Y \in S$ such that $S$ closes via $Y$ with weight $k$. And define "$S$ closes" to mean that there is some $k$ such that $S$ closes with weight $k$. Now define $C$ to be an *abstract Gentzen relation* if the following conditions hold:

$P_1$: The property "$S$ does not close" is an analytic consistency property—or equivalently, the property "$S$ closes" is an analytic inconsistency property. In detail:

(a) If $\{S,\alpha,\alpha_1\}$ or $\{S,\alpha,\alpha_2\}$ closes, so does $\{S,\alpha\}$.

(b) If $\{S,\beta,\beta_1\}$ and $\{S,\beta,\beta_2\}$ both close, so does $\{S,\beta\}$.

(c) If $\{S,\gamma,\gamma(a)\}$ closes, so does $\{S,\gamma\}$.

(d) If $\{S,\delta,\delta(a)\}$ closes for $a$ new to $\{S,\delta\}$ then $\{S,\delta\}$ closes.

$P_2$: (a) If $\{S,\alpha\}$ closes via $\alpha$ with weight $k$, then either $\{S,\alpha,\alpha_1\}$ or $\{S,\alpha,\alpha_2\}$ closes with weight $<k$.

(b) If $\{S,\beta\}$ closes via $\beta$ with weight $k$, then both $\{S,\beta,\beta_1\}$ and $\{S,\beta,\beta_2\}$ close with weight $<k$.

(c) If $\{S,\gamma\}$ closes via $\gamma$ with weight $k$, then for some parameter $a$, $\{S,\gamma,\gamma(a)\}$ closes with weight $<k$.

(d) If $\{S,\delta\}$ closes via $\delta$ with weight $k$, then for some parameter $a$ which is new to $\{S,\delta\}$, the set $\{S,\delta,\delta(a)\}$ closes with weight $<k$.

$P_3$: If $S$ closes with weight $k$, then any (finite) superset of $S$ closes with weight $k$.

$P_4$: If $\{S,\delta(a)\}$ closes with weight $k$, where $a$ is new to $\{S,\delta\}$ then for *every* parameter $b$, $\{S,\delta(b)\}$ closes with weight $k$.

Before we proceed further with our abstract development, let us stop and verify that $P_1$–$P_4$ all hold for the tableau interpretation of "$S$ closes via $Y$ with weight $k$". Well, we already know $P_1$. Condition $P_2$ is really quite obvious—e.g. (a) holds, because if $\{S,\alpha\}$ closes via $\alpha$ with weight $k$, then there is a closed tableau $\mathcal{T}$ for $S$, of weight $k$, and $\alpha$ is the element of $S$ first used in the construction of $\mathcal{T}$. Then $\alpha$ was used either to adjoin $\alpha_1$ or $\alpha_2$—let us assume it was $\alpha_1$. Then $\mathcal{T}$ is also a tableau for $\{S,\alpha,\alpha_1\}$ of weight $k-1$. Similarly, we can verify (b), (c), (d) of $P_2$.

As for $P_3$, let us remark that if $\mathcal{T}$ is a tableau for $S$, and if $S'$ is a superset of $S$, then $\mathcal{T}$ is *not* necessarily a tableau for $S'$ (because we may in $\mathcal{T}$ have used Rule D to introduce a parameter which though new to $S$ may not be new to $S'$). But given a closed tableau $\mathcal{T}$ for $S$ of weight $k$, we can obviously construct a closed tableau $\mathcal{T}'$ for $S$ of weight $k$ in which all newly introduced parameters lie outside of $S'$ (since $S'$ is finite). Then $\mathcal{T}'$ will be a closed tableau for $S'$ of weight $k$.

As for $P_4$, the verification is a bit more delicate. Suppose $\{S,\delta(a)\}$ closes with weight $k$. Then we construct a closed tableau $\mathcal{T}$ for $\{S,\delta(a)\}$ of weight $k$ in which all newly introduced parameters are distinct from both $a$ and $b$. In this tableau, for every sentence $X$, let $X_b^a$ be the result of substituting $b$ for every occurrence of $a$ in $X$ and let $\mathcal{T}_b^a$ be the result of replacing each point $X$ of $\mathcal{T}$ by $X_b^a$. We assert that $\mathcal{T}_b^a$ is a closed tableau (cf. remarks below). Furthermore, for each $X \in S$, $X_b^a = X$ (because $a$ does not occur in $S$) and $[\delta(a)]_b^a = \delta(b)$ (because $a$ does not occur in $\delta$), hence $\mathcal{T}_b^a$ is a closed tableau for $\{S, \delta(b)\}$.

**Remarks.** A detailed verification of our assertion can be helped by first verifying the following:

For any $X$, let $X' = X_b^a$. Then for any $\alpha, \beta, \gamma, \delta$:

(a)    $(\alpha')_i = (\alpha_i)'$    $[i = 1, 2]$

(b)    $(\beta')_i = (\beta_i)'$    $[i = 1, 2]$

(c)    For $c \neq a$, $[\gamma(c)]' = \gamma'(c)$, but $[\gamma(a)]' = \gamma'(b)$.

(d)    For $c \neq a$, $[\delta(c)]' = \delta'(c)$, but $[\delta(a)]' = \delta'(b)$.

Then it is easily verified that if a point $Y$ of $\mathcal{T}$ comes from $X$ by Rule $A, B, C, D$, then $Y'$ legitimately comes from $X'$ in $\mathcal{T}_b^a$ by an application of the same rule.

Now we return to our abstract development. As with the special case of tableaux, we call $X$ *eliminable* if for *every* set $S$, if $\{S, X\}$ and $\{S, \bar{X}\}$ both close, then $S$ closes. We wish to prove:

**Theorem 1.** *(An abstract form of Gentzen's Hauptsatz). Every $X$ is eliminable.*

This theorem will follow from the following "auxiliary" theorem by a sort of "double induction" argument. First of all, we shall say of two sets $S_1, S_2$ that they close with *combined* weight $k$ if there are integers $k_1, k_2$ such that $S_1$ closes with weight $k_1$, $S_2$ closes with weight $k_2$, and $k_1 + k_2 = k$. Now we call $X$ *k-eliminable* if for every (finite) set $S$, if $\{S, X\}$ and $\{S, \bar{X}\}$ close with combined weight $k$, then $S$ closes. To say that $X$ is eliminable is to say that for every $k$, $X$ is $k$-eliminable. So our task is to show that for every $X$ and every $k$, $X$ is $k$-eliminable. This is done by a double induction on the *degree $n$* of $X$ and the integer $k$. More specifically we show

**Auxiliary Theorem.** *Let $n, k$ be integers with the following 2 properties:*

$C_1$: Every $Y$ of degree $< n$ is eliminable.

$C_2$: For every $Y$ of degree $n$, and for every $k' < k$, $Y$ is $k'$-eliminable. *Then every $X$ of degree $n$ is $k$-eliminable.*

**Proof of Auxiliary Theorem.** Assume that $n, k$ satisfy conditions $C_1$, $C_2$. Let $X$ be of degree $n$, and let $\{S, X\}$ close with weight $k_1$ and $\{S, \bar{X}\}$ close with weight $k_2$, where $k_1 + k_2 = k$. We must show that $S$ closes.

We first make the preliminary observation that if either $k_1$ or $k_2 = 0$, then $S$ must certainly close. For suppose $k_1 = 0$. Then $\{S, X\}$ is already closed. Then for some $Y$, $\{S, X\}$ contains both $Y$, $\bar{Y}$. If $Y \neq X$, then $Y$, $\bar{Y}$ both belong to $S$, so $S$ is closed (and hence closes). If $Y = X$, then $\bar{X} \in S$, hence $\{S, \bar{X}\} = S$, so $S$ closes. Similarly if $k_2 = 0$, then $S$ closes.

So we now assume that $k_1 \geqslant 1$, $k_2 \geqslant 1$. Then for some $Y \in \{S, X\}$, $\{S, X\}$ closes via $Y$ with weight $k_1$ and for some $Z \in \{S, \bar{X}\}$, $\{S, \bar{X}\}$ closes via $Z$ with weight $k_2$. There are now two main cases to consider: $(A)$ $Y \neq X$ or $Z \neq \bar{X}$; $(B)$ $Y = X$ and $Z = \bar{X}$.

*Case A.* We assume $Y \neq X$ (the case $Z \neq \bar{X}$ is obviously handled similarly). Now, $\{S\} = \{S, Y\}$ and so $\{S, Y, X\}$ closes via $Y$ with weight $k_1$. Now we appeal to property $P_2$ (of the definition of an abstract Gentzen relation). If $Y$ is either some $\alpha, \gamma$ or $\delta$, then by (a), (c), (d) of $P_2$ there is some $Y_1$ such that $\{S, X, Y, Y_1\}$ closes with weight $< k_1$, and furthermore if $Y = \alpha$, then $Y_1 = \alpha_1$ or $Y_1 = \alpha_2$; if $Y = \gamma$, then for some $a$, $Y_1 = \gamma(a)$; if $Y = \delta$, then for some $a$ new to $\{S, X, Y\}$, $Y_1 = \delta(a)$. Since $\{S, \bar{X}\}$ closes with weight $k_2$, then $\{S, Y, \bar{X}\}$ closes with weight $k_2$ (because $Y \in S$), hence $\{S, Y, Y_1, \bar{X}\}$ closes with weight $k_2$ (by $P_3$). Thus $\{S, Y, Y_1, X\}$ and $\{S, Y, Y_1, \bar{X}\}$ close with combined weight $< k$, hence by condition $C_2$, $\{S, Y, Y_1\}$ closes. Therefore $\{S, Y\}$ closes (by $P_1$), thus $S$ closes.

If $Y$ is some $\beta$, then $\{S, X, \beta, \beta_1\}$ and $\{S, X, \beta, \beta_2\}$ each closes with weight $< k_1$. Also $\{S, \bar{X}, \beta\}$ closes with weight $k_2$, so $\{S, \bar{X}, \beta, \beta_1\}$ and $\{S, \bar{X}, \beta, \beta_2\}$ each closes with weight $k_2$ (by $P_3$). Thus $\{S, \beta, \beta_1, X\}$ and $\{S, \beta, \beta_1, \bar{X}\}$ close with combined weight $< k$, hence $\{S, \beta, \beta_1\}$ closes. Likewise $\{S, \beta, \beta_2, X\}$ and $\{S, \beta, \beta_2, \bar{X}\}$ close with combined weight $< k$, so $\{S, \beta, \beta_2\}$ closes. Hence $\{S, \beta, \beta_1\}$ and $\{S, \beta, \beta_2\}$ both close, so $\{S, \beta\}$ closes (by $P_1$)—i.e. $S$ closes. This concludes Case A.

*Case B.* $Y = X$ and $Z = \bar{X}$. Then $\{S, X\}$ closes via $X$ with weight $k_1$ and $\{S, \bar{X}\}$ closes via $\bar{X}$ with weight $k_2$. Now one of $X$, $\bar{X}$ is some $\alpha$ or some $\gamma$, and the other is respectively some $\beta$ or some $\delta$. We will assume it is $X$ that is some $\alpha$ or some $\gamma$ (the case that $\bar{X}$ is some $\alpha$ or some $\gamma$ is obviously handled similarly).

Suppose that $X$ is some $\alpha$. Then we have that $\{S, \alpha\}$ closed via $\alpha$ with weight $k_1$ and $\{S, \bar{\alpha}\}$ closed via $\bar{\alpha}$ with weight $k_2$. By $P_2 - (a)$, one of the sets $\{S, \alpha, \alpha_1\}$, $\{S, \alpha, \alpha_2\}$ closed with weight $< k_1$. We shall assume it is $\{S, \alpha, \alpha_1\}$ which closed with weight $< k_1$ (the case that it is $\{S, \alpha, \alpha_2\}$ is obviously handled similarly). Since $\{S, \bar{\alpha}\}$ closes with weight $k_2$, so does $\{S, \bar{\alpha}, \alpha_1\}$ (by $P_3$). Hence $\{S, \alpha_1, \alpha\}$ and $\{S, \alpha_1, \bar{\alpha}\}$ close with combined weight $< k$, therefore $\{S, \alpha_1\}$ closes (by hypothesis $C_2$). Also, since $\{S, \bar{\alpha}\}$ closes with weight $k_2$, then $\{S, \bar{\alpha}, \bar{\alpha}_1\}$ (as well as $\{S, \bar{\alpha}, \bar{\alpha}_2\}$) closes with weight $< k_2$ (by $P_2 - (b)$, since $\bar{\alpha}$ is some $\beta$). And $\{S, \alpha, \bar{\alpha}_1\}$ closes with weight $k_1$ (by $P_3$, since $\{S, \alpha\}$ closes with weight $k_1$), hence $\{S, \bar{\alpha}_1, \alpha\}$ and $\{S, \bar{\alpha}_1, \bar{\alpha}\}$ close with combined weight $< k$. Therefore (again by $C_2$) $\{S, \bar{\alpha}_1\}$

closes. So $\{S,\alpha_1\}$ and $\{S,\bar{\alpha}_1\}$ both close. But $\alpha_1$ is of degree $<n$, therefore by hypothesis $C_1$, $S$ closes.

It remains to consider the case that $X$ is some $\gamma$. Then $\{S,\gamma\}$ closes via $\gamma$ with weight $k_1$ and $\{S,\bar{\gamma}\}$ closes via $\bar{\gamma}$ with weight $k_2$. By $P_2$, there is some parameter $b$ such that $\{S,\gamma,\gamma(b)\}$ closes with weight $<k_1$. Since $\{S,\bar{\gamma}\}$ closes with weight $k_2$, then $\{S,\bar{\gamma},\gamma(b)\}$ closes with weight $k_2$. Therefore $\{S,\gamma(b),\gamma\}$ and $\{S,\gamma(b),\bar{\gamma}\}$ close with combined weight $<k$, hence $\{S,\gamma(b)\}$ closes (by $C_2$). Also $\{S,\bar{\gamma}\}$ closes via $\bar{\gamma}$ with weight $k_2$, so by $P_2$, $\{S,\bar{\gamma},\bar{\gamma}(a)\}$ closes with weight $k_2$, for some $a$ new to $\{S,\bar{\gamma}\}$. Then by $P_4$, for any parameter $c$, $\{S,\bar{\gamma},\bar{\gamma}(c)\}$ closes with weight $<k_2$. In particular, $\{S,\bar{\gamma},\bar{\gamma}(b)\}$ closes with weight $<k_2$. Also $\{S,\gamma,\bar{\gamma}(b)\}$ closes with weight $k_1$ (by $P_3$), since $\{S,\gamma\}$ closes with weight $k_1$. So $\{S,\bar{\gamma}(b),\gamma\}$ and $\{S,\bar{\gamma}(b),\bar{\gamma}\}$ close with combined weight $<k$, hence $\{S,\bar{\gamma}(b)\}$ closes (by $C_2$). Thus $\{S,\gamma(b)\}$ and $\{S,\bar{\gamma}(b)\}$ both close, hence $S$ closes (by hypothesis $C_1$, because $\gamma(b)$ is of degree $<n$). This concludes the proof of the auxiliary theorem.

**Proof of Hauptsatz (Theorem I).** Suppose some $X$ were non-eliminable. Then there would have to be a smallest integer $n$ such that some $X$ of degree $n$ were non-eliminable. Then there would have to be a smallest integer $k$ such that some $X$ of degree $n$ is not $k$-eliminable. We then have: (1) every $X$ of degree $<n$ is eliminable; (2) every $X$ of degree $n$ is $k'$-eliminable for all $k'<k$. Then by the auxiliary theorem, every $X$ of degree $n$ is $k$-eliminable. This contradicts the assertion that some $X$ of degree $n$ is not $k$-eliminable.

## § 3. Some Applications of the Hauptsatz

**Corollary 1.** *If* $\{S_1,X\}$ *and* $\{S_2,\bar{X}\}$ *both close, then* $S_1 \cup S_2$ *closes.*

**Proof.** Suppose $\{S_1,X\}$ and $\{S_2,\bar{X}\}$ both close. Since $\{S_1,X\}$ closes, so does $\{S_1,S_2,X\}$ (by $P_3$). Since $\{S_2,\bar{X}\}$ closes, so does $\{S_1,S_2,\bar{X}\}$ (again by $P_3$). Then $\{S_1,S_2\}$ closes by Theorem I.

Next we obtain a wholly *constructive* proof of

**Corollary 2.** *The set of (unsigned) sentences provable by tableaux is closed under modus ponens—i.e. if* $X$, $X \supset Y$ *are both provable, so is* $Y$.

**Proof.** Suppose $X$, $X \supset Y$ are both provable. Then $\{FX\}$ and $\{FX \supset Y\}$ both close. Since $\{FX \supset Y\}$ closes, so does $\{FY,FX \supset Y\}$. Since $\{FX\}$ closes, so does $\{FY,TX \supset Y,FX\}$. Also $\{FY,TX \supset Y,TY\}$ closes (because it is closed). Therefore $\{FY,TX \supset Y\}$ closes (by $P_1 - (b)$, taking $TX \supset Y$ for $\beta$). So $\{FY,FX \supset Y\}$ and $\{FY,TX \supset Y\}$ both close, hence by the Hauptsatz, $FY$ closes—i.e. $Y$ is provable.

N.B.: Corollary 2 should not be confused with the relatively trivial fact (discussed in Chapter VII) that the addition of modus ponens to the tableau rules does not increase the class of provable formulas.

The next 3 corollaries are more in the character of lemmas.

**Corollary 3.** *(a) If* $\{S,\alpha\}$ *closes, so does* $\{S,\alpha_1,\alpha_2\}$. *(b) If* $\{S,\beta\}$ *closes, then* $\{S,\beta_1\}$ *and* $\{S,\beta_2\}$ *both close.*

**Proof.** (a) Suppose $\{S,\alpha\}$ closes. Then $\{S,\alpha,\alpha_1,\alpha_2\}$ closes. Now, $\{S,\bar{\alpha},\alpha_1,\alpha_2,\bar{\alpha}_1\}$ and $\{S,\bar{\alpha},\alpha_1,\alpha_2,\bar{\alpha}_2\}$ are both closed, hence they both close. Therefore $\{S,\bar{\alpha},\alpha_1,\alpha_2\}$ closes (by $P_1-(b)$). So $\{S,\alpha_1,\alpha_2,\alpha\}$ and $\{S,\alpha_1,\alpha_2,\bar{\alpha}\}$ both close, so $\{S,\alpha_1,\alpha_2\}$ closes by the Hauptsatz.

(b) Suppose $\{S,\beta\}$ closes. Then $\{S,\beta,\beta_1\}$ closes. Now $\{S,\beta_1,\bar{\beta},\bar{\beta}_1\}$ is closed, hence closes, so $\{S,\beta_1,\bar{\beta}\}$ closes (by $P_1-(a)$). Thus $\{S,\beta_1,\beta\}$ and $\{S,\beta_1,\bar{\beta}\}$ both close, so $\{S,\beta_1\}$ closes (by the Hauptsatz). A similar argument shows that $\{S,\beta_2\}$ closes.

**Corollary 4.** *(a) If* $\{S,\bar{\gamma}\}$ *and* $\{S,\gamma(a)\}$ *both close, then S closes. (b) If* $\{S,\bar{\delta}\}$ *and* $\{S,\delta(a)\}$ *both close, and if a is new to* $\{S,\delta\}$ *then S closes.*

**Proof.** (a) Suppose $\{S,\bar{\gamma}\}$ and $\{S,\gamma(a)\}$ both close. Since $\{S,\gamma(a)\}$ closes, so does $\{S,\gamma\}$ (because since $\{S,\gamma(a)\}$ closes, then $\{S,\gamma,\gamma(a)\}$ closes, hence $\{S,\gamma\}$ closes by $P_1-(c)$). Thus $\{S,\bar{\gamma}\}$ and $\{S,\gamma\}$ both close, so S closes.

(b) Proof like (a), using $P_1-(d)$ in place of $P_1-(c)$.

**Corollary 5.** (a) *If* $\{S,\gamma \supset \gamma(a)\}$ *closes, so does S.*

(b) *If* $\{S,\delta \supset \delta(a)\}$ *closes, and if a is new to* $\{S,\delta\}$, *then S closes.*

**Proof.** (a) If $\{S,\gamma \supset \gamma(a)\}$ closes, then by Corollary 3–(b), $\{S,\bar{\gamma}\}$ and $\{S,\gamma(a)\}$ both close. The result then follows by Corollary 4–(a).

(b) Proof like (a), using Corollary 4–(b) in place of Corollary 4–(a).

From Corollary 5 follows by an obvious induction:

**Theorem 2.** *If R is finite and regular and if no critical parameter of R occurs in S and if R ∪ S closes, then S closes.*

In Chapter VII we gave a wholly constructive method of obtaining an associate R of S given a closed tableau for S. Now we can do the converse.

**Theorem 3.** *If S has an associate, then S closes.*

**Proof.** Let R be an associate of S. Then $R \cup S$ is *truth-functionally* unsatisfiable, hence there is a closed tableau for $R \cup S$ using only Rules A, B. (This by the completeness theorem for propositional logic, which, though semantical, is wholly constructive.) Then S closes by Theorem 2.

We now know by purely constructive arguments that S closes iff S has an associate. Thus we now have

**Theorem 4.** *(A syntactic form of Herbrand's Theorem)* (a) *If* $\{S, X\}$ *and* $\{S, \bar{X}\}$ *both have associates, then S has an associate.*

(b) *The set of all X such that* $\{\sim X\}$ *has an associate is closed under modus ponens.*

## Chapter XIII

# Prenex Tableaux

## § 1. Prenex Formulas

By a *prenex* formula is meant a formula of the form $(q_1 x_1) \cdots (q_n x_n)(M)$, where each $q_i$ is one of the quantifier symbols "$\forall$", "$\exists$", and $x_i \neq x_j$, for $i \neq j$, and $M$ is a quantifier-free formula (i.e. $M$ contains no quantifiers at all). One sometimes refers to $M$ as the *matrix* of the prenex formula $(q_1 x_1) \cdots (q_n x_n)(M)$, and $(q_1 x_1) \cdots (q_n x_n)$ is called the *prefix*.

It is a well-known result of Quantification Theory that any formula $X$ can be put into prenex normal form—i.e. $X$ is equivalent to some prenex formula $Y$. The proof is based on the following basic equivalences (in these equivalences, $\varphi(x)$ is any formula, $\psi$ is a formula, $y$ is a variable which has no free occurrence in $\varphi(x)$ or $\psi$, and $\varphi(y)$ is the result of substituting $y$ for all occurrences of $x$ in $\varphi(x)$):

$$\sim(\forall x)\varphi(x) \simeq (\exists x)\sim\varphi(x)$$
$$\sim(\exists x)\varphi(x) \simeq (\forall x)\sim\varphi(x)$$
$$(\forall x)\varphi(x) \wedge \psi \simeq (\forall y)[\varphi(y) \wedge \psi]$$
$$\psi \wedge (\forall x)\varphi(x) \simeq (\forall y)[\psi \wedge \varphi(y)]$$
$$(\exists x)\varphi(x) \wedge \psi \simeq (\exists y)[\varphi(y) \wedge \psi]$$
$$\psi \wedge (\exists x)\varphi(x) \simeq (\exists y)[\psi \wedge \varphi(y)]$$
$$(\forall x)\varphi(x) \vee \psi \simeq (\forall y)[\varphi(y) \vee \psi]$$
$$\psi \vee (\forall x)\varphi(x) \simeq (\forall y)[\psi \vee \varphi(y)]$$
$$(\exists x)\varphi(x) \vee \psi \simeq (\exists y)[\varphi(y) \vee \psi]$$
$$\psi \vee (\exists x)\varphi(x) \simeq (\exists y)[\psi \vee \varphi(y)]$$
$$\psi \supset (\exists x)\varphi(x) \simeq (\exists y)[\psi \supset \varphi(y)]$$
$$(\exists x)\varphi(x) \supset \psi \simeq (\forall y)[\varphi(y) \supset \psi]$$
$$\psi \supset (\forall x)\varphi(x) \simeq (\forall y)[\psi \supset \varphi(y)]$$
$$(\forall x)\varphi(x) \supset \psi \simeq (\exists y)[\varphi(y) \supset \psi]$$

These equivalences enable us to move interior quantifiers to the front of a formula. A complete proof that any formula can be put into prenex form can be found in virtually any introductory text (e.g. Church [1], Kleene [1], Mendelson [1]), though the reader not familiar with the proof should have no difficulty working one out using the exercises below.

*Exercise 1.* Suppose $A$, $B$ are respectively equivalent to $A_1$, $B_1$. Prove:

(1)  $\sim A$ is equivalent to $\sim A_1$.
(2)  $(qx)A$ is equivalent to $(qx)A_1$ (where $q$ is $\forall$ or $\exists$).
(3)  For each of the binary connectives $\circ$, $A \circ B$ is equivalent to $A_1 \circ B_1$.
(4)  $(qx)A$ is equivalent to $A$, if $x$ has no free occurrence in $A$.

*Exercise 2.* Suppose $A$, $B$ are prenex formulas. Using Exercise 1 and the basic equivalences given previously, show by induction on the number of quantifiers in the prefixes of $A$, $B$, that:

(1)  $\sim A$ can be put into prenex form.
(2)  $A \circ B$ can be put into prenex form.
(3)  $(\forall x)A$ and $(\exists x)A$ can be put into prenex form.

*Exercise 3.* Using Exercise 2, show that any formula $A$ can be put into prenex form (use induction on the number of logical connectives and quantifiers in $A$).

### § 2.  Prenex Tableaux

We now describe a proof procedure[1]) for *prenex* sentences, which is like tableaux, except that we need no branching!

Let $S$ be a finite set of prenex sentences. By a *prenex-tableau* for $S$ we mean a tableau for $S$ which uses only the quantificational rules $C$, $D$. Since we do not use rule $B$, then naturally a prenex tableau has only one branch.

We call a prenex tableau *P-closed* if the set of formulas of (the one branch of) the tableau is *truth-functionally* unsatisfiable.

Of course, given a $P$-closed prenex tableau for $S$, we can further extend it—using just the truth-functional rules $A$, $B$—to an ordinary tableau which is closed (in the ordinary sense). (This follows from the completeness theorem for tableaux for propositional logic.) We thus have

**Theorem 1.** *Given a P-closed prenex tableau for $S$, we can construct a closed tableaux for $S$ in which all applications of the quantificational rules precede all applications of the truth-functional rules.*

Our main theorem now is

**Theorem 2.** *(Completeness theorem for prenex tableaux). If a finite set $S$ of prenex sentences is unsatisfiable, then there exists a P-closed prenex tableau for $S$.*

---

[1]) This procedure is essentially that given in the Appendix to Quine [1].

To prove Theorem 2, we modify (in the obvious manner) our "systematic" construction of tableaux. Specifically, we start our tableau with the elements of $S$ (in any order). This concludes the 0-th stage. Now suppose we have completed the $n$-th stage. If the tableau at hand is already $P$-closed, or if all lines which are $\gamma$'s or $\delta$'s have been used, then we stop. Otherwise we take the highest unused line which is a $\gamma$ or $\delta$, and use it in the same manner as in the systematic procedure for ordinary tableaux. This concludes stage $n + 1$.

Suppose the tableau runs on infinitely without $P$-closing. We wish to show that the set $K$ of sentences on the tableau is satisfiable. Clearly $K$ has the following properties:

$P_0$: $K$ is truth-functionally consistent (i.e. every finite subset of $K$ is truth-functionally satisfiable).

$P_1$: If $\gamma \in K$, then for every parameter $a$, $\gamma(a) \in K$.

$P_2$: If $\delta \in K$, then for at least one parameter $a$, $\delta(a) \in K$.

We might call a set $K$ having properties $P_0$, $P_1$, $P_2$ a $P$-Hintikka set. (Note that $P_1$, $P_2$ are respectively conditions $H_3$, $H_4$ and $P_0$ is a strengthening of condition $H_0$ defining a Hintikka set.) We now obviously need:

**Lemma.** *(Analogue for prenex sentences of Hintikka's lemma). Every $P$-Hintikka set is (first-order) satisfiable.*

We first explicitly state and prove:

**Sub-lemma.** *Every quantifier-free set $M$ which is truth-functionally satisfiable is first-order satisfiable.*

**Proof of Sub-lemma.** Let $v$ be a Boolean valuation which satisfies $M$. Let $v_0$ be the restriction of $v$ to all *atomic* sentences (thus $v_0$ is an atomic valuation). We know that every atomic valuation can be uniquely extended to a first-order valuation, so let $v'$ be the first-order valuation which extends $v_0$. Then $v'$ is also a Boolean valuation, and $v'$ agrees with $v$ on the set $M_0$ of all *atomic* subformulas of elements of $M$. Therefore $v'$ agrees with $v$ on $M$ (this by induction, since Boolean valuations agreeing on $X$ agree on $\sim X$, and 2 Boolean valuations agreeing on $X$ and $Y$ must also agree on $X \wedge Y$, $X \vee Y$, and $X \supset Y$). Thus all elements of $M$ are true under the *first-order* valuation $v'$, so $M$ is first-order satisfiable.

**Proof of Lemma.** Let $K$ be a $P$-Hintikka set. Let $K_0$ be the set of elements of $K$ which contain no quantifiers. By hypothesis $P_0$, all finite subsets of $K$ are truth-functionally satisfiable, so obviously all finite subsets of $K_0$ are truth-functionally satisfiable. Thus by the Compactness Theorem for propositional logic, $K_0$ is truth-functionally satisfiable.

Then by the sub-lemma, $K_0$ is first-order satisfiable. Let $v$ be a first-order valuation which satisfies $K_0$. We assert that every element $X$ of $K$ is true under $v$. We prove this by induction on the number $n$ of quantifiers of $X$.

If $n=0$, then $X \in K_0$, hence $X$ is true under $v$.

Now suppose that every element of $K$ with $(n-1)$ quantifiers is true under $v$. (a) Suppose $\gamma \in K$ and $\gamma$ has $n$ quantifiers. Then $\gamma(a_1)$, $\gamma(a_2)$,..., $\gamma(a_i)$,..., are all in $K$, and they all have $(n-1)$ quantifiers. Then by the inductive hypothesis, $\gamma(a_1)$,..., $\gamma(a_i)$,..., are all true under $v$. Hence $\gamma$ is true under $v$. (b) Suppose $\delta \in K$ and $\delta$ has $n$ quantifiers. Then for at least one parameter $a$, $\delta(a) \in K$; also $\delta(a)$ has $(n-1)$ quantifiers. Then by the inductive hypothesis, $\delta(a)$ is true under $v$. Hence $X$ is true under $v$. This concludes the proof.

We have now shown that if a systematic $P$-tableau for $S$ is infinite, then $S$ is satisfiable. Suppose now that a systematic $P$-tableau for $S$ terminates without $P$-closing (which, incidentally, can happen only if all quantifiers of the prefix are existential). Then $S$ is satisfiable in the finite domain $\{a_1,...,a_n\}$ of the parameters which were introduced (this by an obvious modification of the proof of the above lemma, which we leave to the reader). Thus if $S$ is unsatisfiable, then the systematic $P$-tableau for $S$ must $P$-close. This proves Theorem 2.

Theorems 1 and 2 at once yield:

**Theorem 3.** *(Semantical version of Gentzen's Extended Hauptsatz modified for tableaux). If a set $S$ of prenex sentences is unsatisfiable, then there exists a closed tableau for $S$ in which all applications of Rules C, D precede all applications of Rules A, B.*

By Theorem 3 and the fact that a closed tableau for $S$ implies the unsatisfiability of $S$, we have

**Theorem 3′.** *(Syntactical version of Gentzen's Extended Hauptsatz modified for tableaux). If there exists a closed tableau for S, where S is a set of prenex sentences, then there exists a closed tableau for S in which all applications of Rules C, D precede all applications of Rules A, B.*

Although Theorem 3′ is purely syntactic, we have proved it by a model-theoretic argument. It can also be proved by a purely syntactic argument.

**Discussion.** It has occurred to us that the $P$-tableau method can be modified in such a manner that the corresponding completeness proof does not require any appeal to a *systematic* tableau. This can be done by incorporating the Henkin-Hasenjaeger idea in the following manner.

Consider the complete regular set $M$ of Chapter IX. Replace rule $D$ by the following rule:

*Rule D':*
$$\frac{\delta_i}{\delta_i(a_i)}$$

Now consider a finite set $S$ of prenex sentences. Let $S^*$ be the *closure* of $S$ under operations $C$ and $D'$—i.e. the intersection of all supersets of $S$ which are closed under applications of rules $C$, $D'$. If $S^*$ is truth-functionally inconsistent, then $S$ is unsatisfiable (why?). If $S^*$ is truth-functionally consistent, then $S^*$ is a $P$-Hintikka set (verify!), in which case $S$ is satisfiable. Therefore $S$ is satisfiable iff $S^*$ is truth-functionally consistent. Thus $S$ is unsatisfiable iff $S$ can be extended to a truth-functionally unsatisfiable set using only *finitely* many applications of Rules $C$, $D'$. Put otherwise, $S$ is unsatisfiable iff there exists a $P$-closed prenex tableau for $S$ using rules $C$, $D'$. This gives a completeness proof for this modified tableau system which does not appeal to any systematic construction.

## Chapter XIV

# More on Gentzen Systems

In § 1 of this chapter, we discuss Gentzen's Extended Hauptsatz. In § 2 we establish a new form of this extension which does not appeal to prenex normal form. In § 3 we consider some variants of Gentzen systems which will play a key role in all 3 subsequent chapters.

## § 1. Gentzen's Extended Hauptsatz

Suppose the sequent $X_1,\ldots,X_n \to Y_1,\ldots,Y_k$ is provable in $\mathscr{G}$, and that $X_1,\ldots,X_n$, $Y_1,\ldots,Y_k$ are *prenex* sentences. Then the set $\{TX_1,\ldots,TX_n, FY_1,\ldots,FY_k\}$ is unsatisfiable. Hence by Theorem 3 of the last chapter, there is a closed analytic tableau for this set in which all applications of the quantificational rules precede all applications of the truth-functional rules. If we translate this tableau into a proof in $\mathscr{G}$ of $X_1,\ldots,X_n \to Y_1,\ldots,Y_k$ (in the manner of Chapter XI), we obtain a proof in which all applications of the truth-functional rules precede all applications of the quantificational rules (we recall that proof trees are displayed upside down!). We thus have

**Theorem 1.** *(Gentzen's Extended Hauptsatz). If* $U \to V$ *is provable in* $\mathscr{G}^*$, *where* $U$, $V$ *are sets of prenex sentences, then there exists a proof of* $U \to V$ *in* $\mathscr{G}$ *in which all applications of the truth-functional rules precede all applications of the quantificational rules.*

## § 2. A New Form of the Extended Hauptsatz

Now we consider a further extension of Gentzen's Extended Hauptsatz which makes no appeal to prenex normal form.

We let $\mathscr{G}'$ be the system obtained from $\mathscr{G}$ by replacing the quantificational rules by the following rules:

$$\mathsf{U}_1' : \quad \frac{U, \varphi(a) \to V \quad U \to V, (\forall x)\varphi(x)}{U \to V},$$

provided $(\forall x)\varphi(x)$ is a subformula of (some term of) $U \to V$

$$\mathsf{U}_2' : \quad \frac{U \to V, \varphi(a) \quad U, (\forall x)\varphi(x) \to V}{U \to V},$$

provided the same as above and also that $a$ does not occur in $U \to V$.

$$\exists_1' : \quad \frac{U \to V, \varphi(a) \quad U, (\exists x)\varphi(x) \to V}{U \to V},$$

provided that $(\exists x)\varphi(x)$ is a subformula of $U \to V$.

$$\exists_2' : \quad \frac{U, \varphi(a) \to V \quad U \to V, (\exists x)\varphi(x)}{U \to V},$$

with the same proviso as $\exists_1'$ and the same proviso as $\mathsf{U}_2'$.

The quantificational rules of $\mathscr{G}'$ are in the spirit of cut rules (combined with the quantificational rules of $\mathscr{G}$), but proofs in $\mathscr{G}'$ nevertheless obey the subformula principle because of the proviso that $(\forall x)\varphi(x)$ (respectively $(\exists x) \varphi(x)$) must be a subformula of $U \to V$.

In uniform notation, the quantificational rules of $\mathscr{G}'$ are as follows:

$$C' : \quad \frac{|S, \gamma(a)| \quad |S, \bar{\gamma}|}{|S|},$$

provided $\gamma$ is a subformula of some element of $S$.

$$D' : \quad \frac{|S, \delta(a)| \quad |S, \bar{\delta}|}{|S|},$$

provided $\delta$ is a subformula of some element of $S$ and $a$ does not occur in any element of $S$.

By a *normal* proof in $\mathscr{G}'$ we shall mean a proof in which all applications of the truth-functional rules precede all applications of the quantificational rules. We wish to prove

**Theorem 2.** *(A new form of Gentzen's Extended Hauptsatz). Every valid sequent $U \to V$ has a normal proof in $\mathscr{G}'$.*

We note that Theorem 2 is asserted for *all* sequents $U \to V$, not just those in which the terms of $U \to V$ are prenex sentences. We will establish Theorem 2 as a consequence of the Fundamental Theorem modified for *signed* formulas as follows.

We now use "$\gamma$" and "$\delta$" for signed formulas, and "$Q$" for either some $\gamma$ or some $\delta$. The notation $Q \supset Q(a)$ now makes no sense, since we do not put logical connectives between signed formulas. We shall however define the signed formula $Q \Rightarrow Q(a)$ as follows:

$$T(\forall x)\varphi(x) \Rightarrow T\,\varphi(a) =_{df} T(\forall x)\varphi(x) \supset \varphi(a)$$
$$F(\forall x)\varphi(x) \Rightarrow F\,\varphi(a) =_{df} T\,\varphi(a) \supset (\forall x)\varphi(x)$$
$$T(\exists x)\varphi(x) \Rightarrow T\,\varphi(a) =_{df} T(\exists x)\varphi(x) \supset \varphi(a)$$
$$F(\exists x)\varphi(x) \Rightarrow F\,\varphi(a) =_{df} T\,\varphi(a) \supset (\exists x)\varphi(x)$$

Our definition of $\Rightarrow$ is reasonable in the sense that under any interpretation, $Q \Rightarrow Q(a)$ is true iff either $Q$ is false or $Q(a)$ is true. Now we define a regular set $R$ of signed formulas in the same manner as for unsigned formulas, only replacing $Q \supset Q(a)$ in the definition by $Q \Rightarrow Q(a)$. And similarly we define an associate $R$ of a set $S$ of signed formulas (only replacing "weak subformula" by "subformula"). Now if $\mathscr{I}$ is a closed analytic tableau for a set $S$ of signed formulas, then the set $R$ of all elements $Q \Rightarrow Q(a)$ such that $Q(a)$ was inferred from $Q$ in $\mathscr{I}$ is an associate of $S$ (the proof is the same as the case for unsigned formulas).

Now for the proof of Theorem 2. Let us first note that a set $\{S, Q \Rightarrow Q(a)\}$ is truth-functionally unsatisfiable iff each of the 2 sets $\{S, \bar{Q}\}$, $\{S, Q(a)\}$ is truth-functionally unsatisfiable; a set $\{S, Q_1 \Rightarrow Q_1(a_1),\ Q_2 \Rightarrow Q_2(a_2)\}$ is truth-functionally unsatisfiable iff each of the 4 sets $\{S, \bar{Q}_1, \bar{Q}_2\}$, $\{S, \bar{Q}_1, Q_2(a_2)\}$, $\{S, Q_1(a_1), \bar{Q}_2\}$, $\{S, Q_1(a_1), Q_2(a_2)\}$ is truth-functionally unsatisfiable—and in general, $\{S, Q_1 \Rightarrow Q_1(a_1), ..., Q_n \Rightarrow Q_n(a_n)\}$ is truth-functionally unsatisfiable iff each of the $2^n$ sets $\{S, A_1, ..., A_n\}$ is unsatisfiable, where $A_1$ is either $\bar{Q}_1$ or $Q_1(a_1)$, $A_2$ is $\bar{Q}_2$ or $Q_2(a_2)$, ..., $A_n$ is either $\bar{Q}_n$ or $Q_n(a_n)$.

Now suppose a sequent $U \to V$ is valid; let $S$ be the set of signed formulas such that $|S|$ is the sequent $U \to V$. Then $S$ is unsatisfiable, so $S$ has an associate $R$. We arrange $R$ is a regular sequence $\langle Q_1 \Rightarrow Q_1(a_1),$ $Q_2 \Rightarrow Q_2(a_2), ..., Q_n \Rightarrow Q_n(a_n) \rangle$. Letting $A_1$ be either $\bar{Q}_1$ or $Q_1(a_1), ..., A_n$ be either $\bar{Q}_n$ or $Q_n(a_n)$, for each of the $2^n$ choices of the sequence

$\langle A_1, A_2, ..., A_n \rangle$, the set $\{S, A_1, ..., A_n\}$ is truth-functionally unsatisfiable, hence the sequent $|S, A_1, ..., A_n|$ is a *tautology*, and hence obviously has a normal proof in $\mathscr{G}'$. This means that for each of the possible $2^{n-1}$ sequences $\langle A_2, ..., A_n \rangle$, both $|S, \bar{Q}_1, A_2, ..., A_n|$ and $|S, Q_1(a_1), A_2, ..., A_n|$ have a normal proof in $\mathscr{G}'$. Then by an application of Rule $C'$ or $D'$ (depending respectively on whether $Q_1$ is some $\gamma$ or some $\delta$) we obtain a normal proof of $|S, A_2, ..., A_n|$. Thus for each of the $2^{n-1}$ choices of $A_2, ..., A_n$, the sequent $|S, A_2, ..., A_n|$ has a normal proof. This means that for each of the $2^{n-2}$ choices of $A_3, ..., A_n$, the sequent $|S, \bar{Q}_2, A_3, ..., A_n|$ and the sequent $|S, Q_2(a_2), A_3, ..., A_n|$ have normal proofs. Another application of either Rule $C'$ or $D'$ gives a normal proof of $|S, A_3, ..., A_n|$. Continuing in this manner, we finally obtain a normal proof of $|S|$.

## § 3. Symmetric Gentzen Systems

In some of the inference rules of $\mathscr{G}$ (more specifically the negation and implication rules) one transfers a formula from one side of the arrow to the other (or more accurately, one incorporates it into some formula on the other side). We shall need some Gentzen-type systems in which this does not happen. As a result, the systems to which we now turn will possess a valuable feature: In any proof of a sequent $U \to V$, for each sequent $U_1 \to V_1$ used in the proof, each term of $U_1$ will be a subformula of some term of $U$, and each term of $V_1$ will be a subformula of some term of $V$. We might refer to this condition as the *2-sided* subformula principle. (By contrast, in a proof of $U \to V$ in $\mathscr{G}$, if $U_1 \to V_1$ is used in the proof, some term of $U_1$ may be a subformula of some term of $V$ but not of $U$, or some term of $V_1$ may be a subformula of some term of $U$ rather than of $V$.)

*The System $\mathscr{S}$.* We shall now consider sequents $K \to L$ in which $K$, $L$ are (finite) sets of *signed* formulas! The postulates of $\mathscr{S}$ (in uniform notation) are as follows:

Axioms:    $K, X \to L, X$

$\qquad\qquad K, X, \bar{X} \to L \qquad$ where $X$ is atomic

$\qquad\qquad K \to L, X, \bar{X}$

Rules:

(A) $\qquad \dfrac{K, \alpha_1, \alpha_2 \to L}{K, \alpha \to L} \qquad\qquad \dfrac{K \to L, \beta_1, \beta_2}{K \to L, \beta}$

(B) $\qquad \dfrac{K, \beta_1 \to L \quad K, \beta_2 \to L}{K, \beta \to L} \qquad \dfrac{K \to L, \alpha_1 \quad K \to L, \alpha_2}{K \to L, \alpha}$

$$(C) \quad \frac{K, \gamma(a) \rightarrow L}{K, \gamma \rightarrow L} \qquad\qquad \frac{K \rightarrow L, \delta(a)}{K \rightarrow L, \delta}$$

$$(D) \quad \frac{K, \delta(a) \rightarrow L}{K, \delta \rightarrow L} \qquad\qquad \frac{K \rightarrow L, \gamma(a)}{K \rightarrow L, \gamma}$$

(provided $a$ does not occur in the conclusion).

We establish the completeness of $\mathscr{S}$ as follows. By a *variant* of a sequent $K \rightarrow L$ we shall mean any sequent obtained from $K \rightarrow L$ by transferring any number of terms from one side of the arrow to the other but at the same time changing their signs. (Thus, e.g. $TX, FY \rightarrow TZ$ is a variant of $FZ, TX \rightarrow TY$, also a variant of $\rightarrow FX, TY, TZ$, also a variant of $TX \rightarrow TY, TZ$.) Clearly if $K' \rightarrow L'$ is a variant of $K \rightarrow L$, then both sequents are *equivalent*—i.e. true under the same interpretations. We note that for a set $S$ of signed formulas, the sequent $S \rightarrow$ is equivalent to the sequent $|S|$ (where terms are *unsigned* formulas). Thus $S \rightarrow$ is valid iff $|S|$ is valid iff the set $S$ is unsatisfiable. By a variant of the set $S$ we shall mean a variant of the sequent $S \rightarrow$, and we shall also refer to any such variant as a variant of $|S|$.

Let us now consider the system $\mathscr{G}$ in uniform notation. We know that $\mathscr{G}$ is complete even restricting the axiom scheme $|S, X, \bar{X}|$ to the case that $X$ is atomic. Now the set of axioms of $\mathscr{S}$ is precisely the set of all variants of all these axioms $|S, X, \bar{X}|$. And in any application of an inference rule of $\mathscr{G}$, if *all* variants of the premises are provable in $\mathscr{S}$, then all variants of the conclusion are provable in $\mathscr{S}$. It follows then by induction that all variants of all theorems of $\mathscr{G}$ are theorems of $\mathscr{S}$. Now suppose that $K \rightarrow L$ is valid. Then the set $K \cup \bar{L}$ is unsatisfiable (where by $\bar{L}$ we mean the set of *conjugates* of the elements of $L$). Then the sequent $|K, \bar{L}|$ (of *unsigned* formulas) is provable in $\mathscr{G}$ (by the completeness of $\mathscr{G}$). But $K \rightarrow L$ is a variant of $|K, \bar{L}|$. Therefore $K \rightarrow L$ is provable in $\mathscr{S}$. Thus $\mathscr{S}$ is complete.

It is, of course, possible to have an alternative version $\mathscr{S}'$ in which the elements of our sequents are *unsigned* formulas—just use the now familiar device of deleting all "$T$"s and replacing all "$F$"s by "$\sim$"s in all terms of the sequents of the axioms and inference rules (and, of course, reinterpret "$\alpha$", "$\beta$", "$\gamma$", "$\delta$" accordingly). Then if a sequent $K \rightarrow L$ is provable in $\mathscr{S}$, its corresponding sequent $U \rightarrow V$ (of unsigned formulas) is provable in $\mathscr{S}'$. This gives the completeness of $\mathscr{S}'$, because if $X_1, \ldots, X_n \rightarrow Y_1, \ldots, Y_k$ is valid, then $TX_1, \ldots, TX_n \rightarrow TY_1, \ldots, TY_k$ is provable in $\mathscr{S}$, hence $X_1, \ldots, X_n \rightarrow Y_1, \ldots, Y_k$ is provable in $\mathscr{S}'$.

*Exercise 1.* Suppose we delete either the second axiom scheme $K, X, \bar{X} \rightarrow L$ or the third axiom scheme $K, \rightarrow L, X, \bar{X}$ and also delete the left halves of rules $A$, $B$, $C$, $D$ or the right halves, but we add the postulate:

$E$:
$$\frac{K \to L}{\bar{L} \to \bar{K}}$$

Show that we obtain an equivalent system.

*The System $\mathscr{S}^*$ of Stroke Sequents.* It will prove convenient to consider another system $\mathscr{S}^*$ even though it is little more than a notational variant of $\mathscr{S}$.

We shall use Sheffer's stroke symbol "$|$" and define a *stroke sequent* as an ordered triple $\langle K, |, L \rangle$—which we write more simply as $K | L$. We define $K | L$ to be *true* under an interpretation if at least one element of $K \cup L$ is false. Stated otherwise, $K | L$ is equivalent to the GENTZEN sequent $K \to \bar{L}$. We now convert our system $\mathscr{S}$ for GENTZEN sequents to a complete system $\mathscr{S}^*$ for stroke sequents by replacing in all postulates all GENTZEN sequents $K \to L$ by the stroke sequent $K | \bar{L}$. In detail, the postulates of $\mathscr{S}^*$ are as follows:

*Axioms.*
$$\left. \begin{array}{l} K, X | L, \bar{X} \\ K, X, \bar{X} | L \\ K | L, X, \bar{X} \end{array} \right\} \quad \text{where } X \text{ is atomic}$$

*Rules.*

$A_1$: $\dfrac{K, \alpha_1, \alpha_2 | L}{K, \alpha | L}$ $\qquad$ $A_2$: $\dfrac{K | L, \alpha_1, \alpha_2}{K | L, \alpha}$

$B_1$: $\dfrac{K, \beta_1 | L \quad K, \beta_2 | L}{K, \beta | L}$ $\qquad$ $B_2$: $\dfrac{K | L, \beta_1 \quad K | L, \beta_2}{K | L, \beta}$

$C_1$: $\dfrac{K, \gamma(a) | L}{K, \gamma | L}$ $\qquad$ $C_2$: $\dfrac{K | L, \gamma(a)}{K | L, \gamma}$

$D_1$: $\dfrac{K, \delta(a) | L}{K, \delta | L}$ $\qquad$ $D_2$: $\dfrac{K | L, \delta(a)}{K | L, \delta}$

(providing $a$ is new to the conclusion)

Two sets are called *incompatible* if their union is unsatisfiable. It is obvious that $K \to \bar{L}$ is provable in $\mathscr{S}$ iff $K | L$ is provable in $\mathscr{S}^*$. Hence if $K$ is incompatible with $L$, then $K \to \bar{L}$ is valid, hence provable in $\mathscr{S}$, and so $K | L$ is provable in $\mathscr{S}^*$. Thus $\mathscr{S}^*$ is complete in the sense that if $K$ is incompatible with $L$ (i.e. if $K | L$ is valid) then $K | L$ is provable in $\mathscr{S}^*$.

The system $\mathscr{S}^*$ is completely symmetric (the system $\mathscr{S}$ might be aptly described as skew-symmetric!). We could, of course, have used just rules $A_1$, $B_1$, $C_1$, $D_1$ (or alternatively $A_2$, $B_2$, $C_2$, $D_2$) and just one of the second and third axiom schemata, had we added the postulate:

$E$:
$$\frac{K \mid L}{L \mid K}.$$

The purpose of introducing the systems $\mathscr{S}, \mathscr{S}^*$ will become apparent in the next three chapters.

## Chapter XV
# Craig's Interpolation Lemma and Beth's Definability Theorem

### § 1. Craig's Interpolation Lemma

A formula $Z$ is called an *interpolation* formula for a formula $X \supset Y$ if all predicates and parameters of $Z$ occur both in $X$ and in $Y$, and if $X \supset Z$, $Z \supset Y$ are both valid. Craig's celebrated Interpolation lemma says that for any valid sentence $X \supset Y$: (*i*) if $X$, $Y$ have at least one predicate in common, then there exists an interpolation sentence for $X \supset Y$; (*ii*) if $X$, $Y$ have no predicates in common, then either $Y$ is valid or $X$ is unsatisfiable.

We will now adjoin the propositional constants $t, f$ to our object language (cf. Exercise 2, end of Chapter I), and so case (*ii*) can be subsumed under case (*i*) as follows. If $Y$ is valid, then $X \supset t$, $t \supset Y$ are both valid, so then $t$ is an interpolation formula for $X \supset Y$. If $X$ is unsatisfiable, then $X \supset f$, $f \supset Y$ are both valid, so then $f$ is an interpolation formula for $X \supset Y$.

There is a corresponding interpolation lemma for propositional logic: If $X \supset Y$ is a tautologous formula of propositional logic, then there exists a formula $Z$ (again called an interpolation formula for $X \supset Y$) of propositional logic such that all *propositional variables* of $Z$ occur in $X$ and in $Y$ and such that $X \supset Z$, $Z \supset Y$ are both tautologies. [For example, $q$ is an interpolation formula for $(p \wedge q) \supset (p \vee q)$.]

Returning to First-Order Logic, a formula $Z$ is called an interpolation formula for a *sequent* $U \rightarrow V$ if every predicate and parameter of $Z$ occurs in at least one element of $U$ and at least one element of $V$ and if both sequents $U \rightarrow Z$, $Z \rightarrow V$ are valid. Obviously $Z$ is an interpolation formula for the *sequent* $X_1, \ldots, X_n \rightarrow Y_1, \ldots, Y_k$ iff $Z$ is an interpolation formula for the *sentence* $(X_1 \wedge \cdots \wedge X_n) \supset (Y_1 \vee \cdots \vee Y_k)$, so the existence of interpolation formulas for all valid *sequents* is equivalent to the existence of interpolation formulas for all valid sentences $X \supset Y$. So we shall consider Craig's lemma in the equivalent form that there exist interpolation formulas for all valid *sequents*.

**Proof of Craig's Lemma.** We can obtain the interpolation lemma as a consequence of our Unifying Principle as follows: Define $\Gamma(S)$ to mean that there exist two sets $S_1$, $S_2$ such that $S = S_1 \cup \bar{S}_2$, and there exists no interpolation formula for the sequent $S_1 \rightarrow S_2$. (Equivalently, $S$ is $\Gamma$-inconsistent if for any two sets $S_1, S_2$ such that $S_1 \cup \bar{S}_2 = S$, there exists an interpolation formula for $S_1 \rightarrow S_2$.) Then Craig's lemma can be obtained by verifying that this $\Gamma$ is an analytic consistency property. However, it is a bit more direct to use the results of Exercise 2, end of Chapter VI, and define $U \vdash V$ to mean that there exists an interpolation formula for the sequent $U \rightarrow V$, and then verify that $\vdash$ is a symmetric GENTZEN relation. This is practically tantamount to taking the symmetric GENTZEN system $\mathscr{S}$ (which we introduced in the last chapter) and showing: (*i*) there exists an interpolation formula for all the axioms; (*ii*) in any application of an inference rule, if there exist interpolation formulas for the premises, then there exists an interpolation formula for the conclusion. This is the course we shall take. Also our proof will be wholly *constructive* in the sense that we shall specify outright interpolation formulas for the axioms, and for any application of an inference rule, given interpolation formulas for the premises, we explicitly exhibit an interpolation formula for the conclusion. Thus given any proof of $U \rightarrow V$ in $\mathscr{S}$, an interpolation formula for $U \rightarrow V$ can be explicitly found. This implies (by earlier results) that given any closed tableau for $U \rightarrow V$, an interpolation formula for $U \rightarrow V$ can be explicitly found.

It makes no great difference whether we work with the system $\mathscr{S}$ (which uses signed formulas) or the system $\mathscr{S}'$ (which uses unsigned formulas) (cf. remark below). Perhaps it might be a bit easier to first work with $\mathscr{S}'$.

Now for the proof. We shall write $U \overset{X}{\rightarrow} V$ to mean that $X$ is an interpolation formula for $U \rightarrow V$.

*Verification for the Axioms.* It is obvious that $X$ is an interpolation formula for $U, X \rightarrow V, X$; $f$ is an interpolation formula for $U, X, \sim X \rightarrow V$; $t$ is an interpolation formula for $U \rightarrow V, X, \sim X$. In other words we have:

(*i*)   $U, X \overset{X}{\rightarrow} V, X$

(*ii*)   $U, X, \sim X \overset{f}{\rightarrow} V$

(*iii*)   $U \overset{t}{\rightarrow} V, X, \sim X$

*Verification for the Truth Functional Rules.*

$A$:   If $U, \alpha_1, \alpha_2 \overset{X}{\rightarrow} V$, then also $U, \alpha \overset{X}{\rightarrow} V$.

   If $U \overset{X}{\rightarrow} V, \beta_1, \beta_2$ then also $U \overset{X}{\rightarrow} V, \beta$.

$B$:   If $U, \beta_1 \overset{X}{\rightarrow} V$ and $U, \beta_2 \overset{Y}{\rightarrow} V$, then $U, \beta \overset{X \vee Y}{\rightarrow} V$.

   If $U \overset{X}{\rightarrow} V, \alpha_1$ and $U \overset{Y}{\rightarrow} V, \alpha_2$, then $U \overset{X \wedge Y}{\rightarrow} V, \alpha$.

Before proceeding further, let us note that we have now proved Craig's lemma for *propositional* logic (because the system $\mathscr{S}'$ without rules $C$, $D$ constitutes a complete system for propositional logic).

*Verification for the Quantificational Rules.*

$C$: Suppose $U, \gamma(a) \xrightarrow{X} V$. Then certainly $U, \gamma \rightarrow X$ and $X \rightarrow V$ are both valid, but $X$ may fail to be an interpolation formula for $U, \gamma \rightarrow V$, because the parameter $a$ might occur in $X$ but not in $U, \gamma$. If $a$ doesn't occur in $X$, then $X$ will be an interpolation formula for $U, \gamma \rightarrow V$. Or if $a$ occurs in $U, \gamma$ then $X$ is still an interpolation formula for $U, \gamma \rightarrow V$. If $a$ occurs in $X$ but not in $U, \gamma$, then take any variable $x$ which does not occur in $X$, and let $X_x^a$ be the result of substituting $x$ for every occurrence of $a$ in $X$. Then $(\forall x) X_x^a$ will be an interpolation formula for $U, \gamma \rightarrow V$. Reason: $(\forall x) X_x^a$ is a sentence $\gamma'$ of type $C$ and $a$ does not occur in $\gamma'$ and $\gamma'(a) = X$. We already know that $U, \gamma \rightarrow \gamma'(a)$ (which is the same as $U, \gamma \rightarrow X$) is valid, but since $a$ does not occur in $U, \gamma, \gamma'$, then $U, \gamma \rightarrow \gamma'$ is valid (indeed $U, \gamma \rightarrow \gamma'$ is derivable from $U, \gamma \rightarrow \gamma'(a)$ by one application of Rule $D$ of our system $\mathscr{S}'$). Thus $U, \gamma \rightarrow \gamma'$ is valid. Also $\gamma'(a) \rightarrow V$ is valid (by hypothesis), so $\gamma' \rightarrow V$ is valid (cf. Rule $C$ of $\mathscr{S}'$, taking $K$ to be empty). Thus, $U, \gamma \rightarrow \gamma'$ and $\gamma' \rightarrow V$ are valid, and certainly all predicates and parameters of $\gamma'$ are both in $U, \gamma$ and in $V$ (because $a$ does not occur in $\gamma'$). Thus $\gamma'$—i.e. $(\forall x) X_x^a$—is an interpolation formula for $U, \gamma \rightarrow V$.

The reader can similarly verify that if $U \xrightarrow{X} V, \delta(a)$, then (i) if $a$ does not occur in $X$, or if $a$ occurs in $X$ and in $V, \delta$, then $U \xrightarrow{X} V, \delta$; (ii) if $a$ occurs in $X$ but not in $V, \delta$, then $U \xrightarrow{(\exists x) X_x^a} V, \delta$.

$D$: Suppose $U, \delta(a) \xrightarrow{X} V$, and that $a$ does not occur in $U, \delta$, $V$. Then $a$ cannot occur in $X$ either, hence $U, \delta \xrightarrow{X} V$.

Likewise if $U \xrightarrow{X} V, \gamma(a)$, then also $U \xrightarrow{X} V, \gamma$. This completes the proof.

**Remark.** If we prefer to work with sequents $U \rightarrow V$ whose terms are *signed* formulas, then the easiest way to modify the above proof is to define an *unsigned* formula $X$ to be an interpolation formula for $U \rightarrow V$ if the signed formula $TX$ is an interpolation formula for $U \rightarrow V$. Then given a proof of $U \rightarrow V$ in the system, we can find an unsigned interpolation formula $X$ for $U$, $V$ exactly as in the proof for $\mathscr{S}'$, and then $TX$ will be a (signed) interpolation formula for $U \rightarrow V$.

Also, if we work with $\mathscr{S}$ rather than $\mathscr{S}'$, we could have halved our labor, appealing to Exercise 1 of the last chapter. Checking for Postulate $E$ is trivial; if $X$ is an interpolation formula for $U \rightarrow V$, then $\sim X$ is an interpolation formula for $\overline{V} \rightarrow \overline{U}$.

*Exercise 1.* Let us define a *Craig sequent* as an ordered triple $(U, X, V)$, where $U$, $V$ are finite sets of sentences and $X$ is a single sentence, and $X$ is an interpolation formula for $U$, $V$. Let us also write $U \rightarrow X \rightarrow V$ for $(U, X, V)$. Now consider the following axiom system for Craig sequents:

*Axioms:*   $U, X \to X \to V, X$
$\qquad\quad\ U, X, {\sim}X \to f \to V$
$\qquad\quad\ U \to t \to V, X, {\sim}X$

*Rules: A:*   $\dfrac{U, \alpha_1, \alpha_2 \to X \to V}{U, \alpha \to X \to V}$ $\qquad\qquad$ $\dfrac{U \to X \to V, \beta_1, \beta_2}{U \to X \to V, \beta}$

$\quad\ B:$ $\dfrac{U, \beta_1 \to X \to V \quad U, \beta_2 \to Y \to V}{U, \beta \to X \vee Y \to V}$ $\qquad$ $\dfrac{U \to X \to V, \alpha_1 \quad U \to Y \to V, \alpha_2}{U \to X \wedge Y \to V, \alpha}$

$\quad\ C:$ $\dfrac{U, \gamma(a) \to X \to V}{U, \gamma \to X \to V}$ $\qquad\qquad$ $\dfrac{U \to X \to V, \delta(a)}{U \to X \to V, \delta}$

provided that either $a$ does not occur in $X$ or that $a$ occurs in $U, \gamma$ or $V, \delta$

$\dfrac{U, \ \gamma(a) \to \gamma'(a) \to V}{U, \ \gamma \to \gamma' \to V}$ $\qquad\qquad$ $\dfrac{U \to S'(a) \to V, S(a)}{U \to S' \to V, S}$

provided that $a$ does          provided that $a$ does
not occur in $U, \gamma, \gamma'$.          not occur in $V, S, S'$.

$\quad\ D:$ $\dfrac{U, \delta(a) \to X \to V}{U, \delta \to X \to V}$ $\qquad\qquad$ $\dfrac{U \to X \to V, \gamma(a)}{U \to X \to V, \gamma}$

provided that $a$ does not occur in the conclusion.

Now define $X$ to be a *special* interpolation formula for $U \to V$ if the Craig sequent $U \to X \to V$ is provable in the above system. Show that if $U \to V$ is valid, then there exists a special interpolation formula for $U \to V$.

*Exercise 2.* The following conditions recursively define the notion of a predicate occurring *positively* in a formula and *negatively* in a formula:

(1) $P$ occurs *positively* in an atomic formula $P a_1, \ldots, a_n$; $P$ occurs *negatively* in ${\sim}P a_1, \ldots, a_n$.

(2) If $P$ occurs positively (negatively) in $\alpha_1$ or in $\alpha_2$, it occurs positively (respectively negatively) in $\alpha$; if $P$ occurs positively (negatively) in $\beta_1$ or in $\beta_2$, it again occurs the same way in $\beta$.

(3) If $P$ occurs positively (negatively) in $\gamma(a)$, it occurs the same way in $\gamma$. Likewise with "$\delta$" in place of "$\gamma$".

It is possible for a predicate to occur both positively and negatively in the same formula—e.g. $P$ occurs both positively and negatively in the formula $(P a \vee {\sim}P b)$.

Now call an interpolation formula $X$ for $U \to V$ a *Lyndon* interpolation formula (for $U \to V$) if every predicate which occurs positively in $X$ occurs positively in both $U$ and $V$, and every predicate which occurs negatively in $X$ occurs negatively in both $U$ and $V$. Prove Roger Lyndon's stronger form of Craig's Interpolation Lemma (sometimes known as the CRAIG-LYNDON interpolation lemma): For any valid sequent $U \to V$ there exists a LYNDON interpolation formula. (Hint: Show that every special interpolation formula for $U \to V$ is a LYNDON interpolation formula for $U \to V$.

*Exercise 3.* An (unsigned) formula $Y$ is said to be in *negation normal form* if "$\supset$" does not occur in $Y$ and if all occurrences of negation signs immediately precede *atomic* formulas. Show that every unsigned formula $X$ is equivalent to a formula $Y$ in negation normal form. Hint: Verify and use the following equivalences:

$$\sim \sim X \simeq X$$
$$\sim(X \wedge Y) \simeq \sim X \vee \sim Y$$
$$\sim(X \vee Y) \simeq \sim X \wedge \sim Y$$
$$X \supset Y \simeq \sim X \vee Y$$
$$\sim(\forall x)X \simeq (\exists x)\sim X$$
$$\sim(\exists x)X \simeq (\forall x)\sim X$$

Now suppose $Y$ is in negation normal form obtained from $X$ using the above equivalences. Prove that a predicate $P$ occurs positively in $X$ iff $P$ has at least one occurrence in $Y$ which is not immediately preceded by a negation sign, and that $P$ occurs negatively in $X$ iff $P$ has at least one occurrence in $Y$ which is immediately preceded by a negation sign.

## § 2. Beth's Definability Theorem

One important use of Craig's Interpolation Lemma is that it yields a remarkably elegant proof of Beth's Definability Theorem which we now discuss.

Consider a set $A$ of closed formulas without parameters. Let us refer to $A$ as a (first-order) *theory*. We refer to the elements of $A$ as the *axioms* of $A$. For the time being, let $A$ be finite. We write $A \vdash X$ to mean that $X$ is true in all interpretations which satisfy $A$, or equivalently that there exists a closed tableau for $\{A, \sim X\}$, or equivalently that the sequent $A \to X$ is provable, or equivalently that if we adjoin the elements of $A$ as additional axioms of the system $Q$, then $X$ becomes provable. If any of these equivalent conditions hold, then $X$ is said to be a *theorem* of the theory $A$.

Now let $P, P_1, \ldots, P_n$ be the predicates which occur in $A$, and let us presently assume that $P$ is of degree one. $P$ is said to be *explicitly* definable from $P_1, \ldots, P_n$ in the theory $A$ if there exists a formula $\varphi(x)$ with just one free variable $x$, whose predicates are all in the set $P_1, \ldots, P_n$ (so $P$ is *not* a predicate of $\varphi(x)$) and such that

$$A \vdash (\forall x)[P(x) \leftrightarrow \varphi(x)],$$

And such a formula $\varphi(x)$ is said to constitute an *explicit* definition of $P$ from $P_1,\ldots,P_n$ in the theory $A$.

Now we say that the axioms of $A$ *implicitly* define $P$ from $P_1,\ldots,P_n$ or that $P$ is *implicitly* definable from $P_1,\ldots,P_n$ in the theory $A$ if the following condition holds: Take a 1-place predicate $P'$ which does not occur in $A$, and let $A'$ be the result of substituting $P'$ for $P$ in every element of $A$. Then $P$ is called implicitly definable from $P_1,\ldots,P_n$ in $A$ if

$$A \cup A' \vdash (\forall x)[P(x) \leftrightarrow P'(x)]$$

Using the completeness theorem, this condition is equivalent to the condition that any two interpretations of $P_1,\ldots,P_n$, $P$ which satisfy $A$ and which agree on $P_1,\ldots,P_n$ must also agree on $P$—or stated otherwise, given any values of $P_1,\ldots,P_n$, there is at most one value of $P$ which satisfies the axioms of $A$. (Why is this equivalent?)

It is obvious that if $A$ defines $P$ explicitly from $P_1,\ldots,P_n$, then it defines $P$ implicitly from $P_1,\ldots,P_n$. For suppose $P$ is explicitly definable from $P_1,\ldots,P_n$ in $A$. Then we have $A \vdash (\forall x)[P(x) \leftrightarrow \varphi(x)]$. Then of course, we also have (for any new predicate $P'$) $A' \vdash (\forall x)[P'(x) \leftrightarrow \varphi(x)]$. Hence $A \cup A' \vdash (\forall x)[P(x) \leftrightarrow \varphi(x)] \wedge (\forall x)[P'(x) \leftrightarrow \varphi(x)]$, hence $A \cup A' \vdash (\forall x)[P(x) \leftrightarrow P'(x)]$ (why?).

Beth's definability theorem says the *converse*—i.e. if $A$ *implicitly* defines $P$ from $P_1,\ldots,P_n$, then $A$ *explicitly* defines $P$ from $P_1,\ldots,P_n$.

Now we shall prove Beth's theorem using Craig's lemma. Without loss of generality, we can assume that $A$ consists of just one sentence (otherwise we can work with the conjunction of the elements of $A$). Thus we are given $A, A' \vdash (\forall x)[P(x) \leftrightarrow P'(x)]$. Since the sequent $A, A' \rightarrow (\forall x)[P(x) \leftrightarrow P'(x)]$ is valid, so is the sequent $A, A' \rightarrow (P(a) \leftrightarrow P'(a))$ (for any parameter $a$). Hence the sequent $A, A' \rightarrow (P(a) \supset P'(a))$ is valid. This sequent is truth-functionally equivalent to the sequent $A, P(a) \rightarrow (A' \supset P'(a))$, hence this latter sequent is valid. In this sequent $P$ does not occur on the right and $P'$ does not occur on the left. Now we take an interpolation formula $X$ for $A, P(a) \rightarrow (A' \supset P'(a))$. Since all predicates of $X$ occur both on the left and right side of the sequent, then $P$ does not occur in $X$—indeed all predicates of $X$ are in the set $P_1,\ldots,P_n$. Also $X$ contains no parameters except (possibly) $a$; let $x$ be a variable new to $X$ and let $\varphi(x) = X_x^a$, so that $\varphi(a) = X$. Then $\varphi(a)$ is an interpolation formula for $A, P(a) \rightarrow (A' \supset P'(a))$. Hence

(1)     $A, P(a) \vdash \varphi(a)$

(2)     $\varphi(a) \vdash (A' \supset P'(a))$.

From (1) we have $A \vdash (P(a) \supset \varphi(a))$. From (2) we have $A' \vdash (\varphi(a) \supset P'(a))$. Hence also $A \vdash (\varphi(a) \supset P(a))$. Hence $A \vdash (P(a) \supset \varphi(a))$ and $A \vdash ((\varphi(a) \supset P(a))$, so $A \vdash (P(a) \leftrightarrow \varphi(a))$. Therefore $A \vdash (\forall x)[P(x) \leftrightarrow \varphi(x)]$, so that $\varphi(x)$ does explicitly define $P$ from $P_1, \ldots, P_n$ in the theory $A$. This concludes the proof of Beth's theorem.

*Exercise.* Show Beth's theorem for the case that $A$ is a denumerable set (interpreting $A \vdash X$ to mean that $X$ is true in all interpretations which satisfy $A$, or equivalently that for some finite subset $A_0$ of $A$, the sequent $A_0 \to X$ is provable).

## Chapter XVI

# Symmetric Completeness Theorems

In this chapter we establish some new and stronger versions of completeness theorems considered earlier. These are closely related to Craig's Interpolation lemma, and they will play a key rôle in our final chapter on linear reasoning.

## § 1. Clashing Tableaux

Suppose we jointly construct a pair $(\mathcal{I}_1, \mathcal{I}_2)$ of analytic tableaux, $\mathcal{I}_1$ for a finite set $S_1$ and $\mathcal{I}_2$ for a finite set $S_2$, alternating between the constructions (though not necessarily methodically—i.e. we may carry out any finite number of steps on one tree, then any finite number of steps on the other, then back to the first, etc.), and respecting Rule D in the sense that if $\delta(a)$ is inferred from $\delta$ in either tree, the parameter $a$ must be new to *both* trees. Call such a pair $(\mathcal{I}_1, \mathcal{I}_2)$ a *joint* pair of tableaux for $S_1, S_2$ or for $S_1 | S_2$.

If the reader would like us to be more precise, then by a joint pair $(\mathcal{I}_1, \mathcal{I}_2)$ for $S_1, S_2$, we mean a pair constructed as follows: we start $\mathcal{I}_1$ with the set $S_1$ (in any order) and $\mathcal{I}_2$ with $S_2$ (in any order). Now suppose $(\mathcal{I}_1, \mathcal{I}_2)$ is a joint pair for $S_1, S_2$ already constructed. Then we may extend the pair by either of the following operations:

(1) If $\mathcal{I}_1'$ is an immediate extension of $\mathcal{I}_1$ by Rule $A, B,$ or $C$, then $(\mathcal{I}_1', \mathcal{I}_2)$ is a joint pair for $S_1, S_2$; if $\mathcal{I}_2'$ is an immediate extension of $\mathcal{I}_2$ by Rule $A, B,$ or $C$, then $(\mathcal{I}_1, \mathcal{I}_2')$ is a joint pair for $S_1, S_2$.

(2) If $\mathcal{I}_1'$ is an immediate extension of $\mathcal{I}_1$ by an application $\dfrac{\delta}{\delta(a)}$ of Rule $D$, and if $a$ is new both to $\mathcal{I}_1$ and to $\mathcal{I}_2$, then $(\mathcal{I}_1', \mathcal{I}_2)$ is a joint pair for $S_1, S_2$.

Likewise, if $\mathcal{I}_2'$ is an immediate extension of $\mathcal{I}_2$ by an application $\dfrac{\delta}{\delta(a)}$ of Rule $D$ and if $a$ is new to both $\mathcal{I}_1$ and $\mathcal{I}_2$.

This defines the notion of a joint pair $(\mathcal{I}_1, \mathcal{I}_2)$ of tableaux for $S_1, S_2$.

We shall say that two sets *clash*—or that one clashes with the other—if their union contains some *atomic* element and its conjugate. We shall say that $\mathscr{I}_1$ *clashes* with $\mathscr{I}_2$—or that $\mathscr{I}_1, \mathscr{I}_2$ *clash*—if *every* branch of $\mathscr{I}_1$ clashes with *every* branch of $\mathscr{I}_2$. And by a *clashing* pair of tableaux for $S_1, S_2$, we mean a joint pair $(\mathscr{I}_1, \mathscr{I}_2)$ of tableaux for $S_1, S_2$ such that $\mathscr{I}_1$ clashes with $\mathscr{I}_2$.

It is obvious that if there exists a clashing pair $(\mathscr{I}_1, \mathscr{I}_2)$ for $S_1, S_2$, then $S_1 \cup S_2$ must be unsatisfiable (why?). We now wish to prove the converse (which is sort of a "double analogue"—or a "symmetric form"—of the completeness theorem)—viz. that if $S_1 \cup S_2$ is unsatisfiable, then there exists a clashing pair $(\mathscr{I}_1, \mathscr{I}_2)$ for $S_1, S_2$.

We could, of course, prove this model-theoretically by making the appropriate changes in our completeness proof for analytic tableaux. More specifically, it should be obvious how to modify our *systematic* constructions of single tableaux to *systematic* constructions of joint pairs of tableaux. Now if we systematically construct a joint pair for $S_1, S_2$ and if after no finite stage do we obtain a clash, then we get two trees $\mathscr{I}_1, \mathscr{I}_2$ (either or both of which may be infinite) and an open branch $B_1$ of $\mathscr{I}_1$ and an open branch $B_2$ of $\mathscr{I}_2$ such that $B_1$ does not clash with $B_2$. Then $B_1 \cup B_2$ will be a Hintikka set, hence $S_1 \cup S_2$ will be satisfiable. Thus if $S_1 \cup S_2$ is unsatisfiable, then a *systematically* constructed pair for $S_1, S_2$ must eventually clash.

To carry out the above proof in detail essentially involves repetition of the labor of our completeness proof for single tableaux. Alternatively, we can obtain our completeness theorem for pairs of tableaux as a consequence of our Unifying Principle as follows: Define $\Gamma(S)$ to mean that there exist sets $S_1, S_2$ such that $S_1 \cup S_2 = S$ and such that there exists *no* clashing pair $(\mathscr{I}_1, \mathscr{I}_2)$ for $S_1, S_2$. Then verify that this $\Gamma$ is an analytic consistency property.

Either of the above methods establishes:

**Theorem 1.** *If $S_1 \cup S_2$ is unsatisfiable, then there exists a clashing pair* $(\mathscr{I}_1, \mathscr{I}_2)$ *for $S_1, S_2$.*

Theorem 1 of course implies

**Theorem 1′.** *If there is a closed tableau $\mathscr{I}$ for $S_1 \cup S_2$, then there is a clashing pair $(\mathscr{I}_1, \mathscr{I}_2)$ for $S_1, S_2$.*

We will shortly discuss a purely syntactic proof of Theorem 1′.

One can of course consider joint pairs of tableaux for propositional logic—just ignore rules $C, D$. Clashing tableaux provide a pleasant proof procedure. We shall consider two examples—the first from propositional logic, the second from Quantification Theory. By a *clashing pair* for a sequent $X_1, ..., X_n \rightarrow Y_1, ..., Y_k$, we shall mean a clashing pair for $(\{TX_1, ..., TX_n\}, \{FY_1, ..., FY_k\})$.

*Example 1.* The following is a clashing pair for $p \supset q, q \supset r \rightarrow p \supset r$:

$$
\begin{array}{ll}
Tp \supset q & Fp \supset r \\
Tq \supset r & Tp \\
Fp \quad | \quad Tq & Fr \\
\quad | \quad Fq \,|\, Tr &
\end{array}
$$

*Example 2.* The following is a clashing pair for the sequent
$(\forall x)(Px \supset Qx), (\forall x)Px \rightarrow (\forall x)Qx$:

| | | | |
|---|---|---|---|
| (1) | $T(\forall x)(Px \supset Qx)$ | (3) | $F(\forall x)Qx$ |
| (2) | $T(\forall x)Px$ | (4) | $FQa$ |
| (5) | $TPa$ | | |
| (6) | $TPa \supset Qa$ | | |

$$(7) \quad FPa \,|\, (8) \quad TQa$$

Note that (8) is the conjugate of (4); (7) is the conjugate of (5).

*Clashing Block Tableaux.* It is obvious how we should define a joint pair $\langle \mathscr{T}_1, \mathscr{T}_2 \rangle$ of block tableaux or of modified block tableaux for $S_1, S_2$, and how to obtain a clashing pair of modified block tableaux for $S_1, S_2$ from a clashing pair of analytic tableaux for $S_1, S_2$. Theorems 1 and 1' thus hold if $\mathscr{I}_1, \mathscr{I}_2, \mathscr{I}$ are modified block tableaux rather than analytic tableaux.

## A Constructive Proof of Theorem 1'.

Though Theorem 1' is purely syntactic, the preceding proof is model-theoretic (since it appeals to the notion of validity). We now wish to sketch an effective procedure for transforming a closed tableau for $S_1 \cup S_2$ into a clashing pair for $S_1, S_2$. To this end, the symmetric system $\mathscr{S}^*$ will be useful. Given a closed tableau for $S_1 \cup S_2$, we know how to obtain a proof of the GENTZEN sequent $S_1 \rightarrow \bar{S}_2$ in $\mathscr{S}$, and hence trivially how to obtain a proof of $S_1 | S_2$ in $\mathscr{S}^*$. The problem then reduces to showing how from a proof of $S_1 | S_2$ in $\mathscr{S}^*$, we can obtain a clashing pair $(\mathscr{I}_1, \mathscr{I}_2)$ of tableaux for $S_1 | S_2$.

This is done by showing that there exists a clashing pair for each axiom of $\mathscr{S}^*$, and in any application of the inference rules, if there are clashing pairs for each premise, then there is a clashing pair for the conclusion. We remark that only the left halves $A_1, B_1, C_1, D_1$ of rules $A, B, C, D$ need be checked, if we use Postulate $E$. And Postulate $E$ can be immediately checked by noting that if $(\mathscr{I}_1, \mathscr{I}_2)$ is a clashing pair for $S_1 | S_2$, then $(\mathscr{I}_2, \mathscr{I}_1)$ is a clashing pair for $S_2 | S_1$. We rely on the reader to check on the details of the proof. Having done this, we obtain (via the system $\mathscr{S}^*$) an effective procedure for obtaining a clashing pair for $S_1, S_2$ given a closed tableau for $S_1 \cup S_2$.

## § 2. Clashing Prenex Tableaux

Let $S_1, S_2$ be finite sets of *prenex* sentences. By a joint pair $(\mathscr{I}_1, \mathscr{I}_2)$ of prenex tableaux for $S_1 | S_2$ we mean a pair of trees constructed using just rules $C$, $D$ of the preceding section. And we say that $\mathscr{I}_1, \mathscr{I}_2$ are *P-clashing* if for the branch $B_1$ of $\mathscr{I}_1$ and the branch $B_2$ of $\mathscr{I}_2$, the set $B_1 \cup B_2$ is *truth-functionally* unsatisfiable.

We leave it to the reader to give an appropriate definition of a *systematic* pair $(\mathscr{I}_1, \mathscr{I}_2)$ for $S_1 | S_2$, and to show that if $(\mathscr{I}_1, \mathscr{I}_2)$ is a completed systematic pair for $S_1 | S_2$ which is not *P*-clashing, then for any branch $B_1$ of $\mathscr{I}_1$ and any branch $B_2$ of $\mathscr{I}_2$ such that $B_1 \cup B_2$ is truth-functionally satisfiable, $B_1 \cup B_2$ is a *P*-Hintikka set. And *P*-Hintikka sets are satisfiable (cf. §2 of Chapter XIII). Thus we have

**Theorem 2.** *If $U | V$ is valid, where $U, V$ are sets of prenex sentences, then there is a P-clashing pair of prenex tableaux for $U | V$.*

We leave it to the reader to show that Theorem 2 implies.

**Theorem 2′**—(A symmetric form of Gentzen's Extended Hauptsatz)— If $U \to V$ is provable in the system $\mathscr{S}$, where $U, V$ are sets of prenex sentences, then $U \to V$ has a proof in $\mathscr{S}$ in which all applications of truth-functional rules precede all applications of quantificational rules.

## § 3. A Symmetric Form of the Fundamental Theorem

We shall call an ordered pair $\langle R_1, R_2 \rangle$ a *joint associate* of an ordered pair $\langle S_1, S_2 \rangle$ if the following conditions hold:
(1) $R_1 \cup R_2$ is an associate of the set $S_1 \cup S_2$.
(2) For each element $Q_1 \supset Q_1(a_1)$ of $R_1, Q_1$ is a weak subformula of some element of $S_1$, and for each element $Q_2 \supset Q_2(a_2)$ of $R_2, Q_2$ is a weak subformula of some element of $S_2$.

We wish to prove:

**Theorem 3.** *(A Symmetric Form of the Fundamental Theorem). If $S_1 \cup S_2$ is unsatisfiable, then there is a joint associate $\langle R_1, R_2 \rangle$ of $\langle S_1, S_2 \rangle$.*

One proves Theorem 3 using the Completeness theorem for clashing tableaux (Theorem 1) in much the same way as one proves the Fundamental Theorem using the Completeness theorem for (single) tableaux. To be specific, let $(\mathscr{I}_1, \mathscr{I}_2)$ be a clashing pair of tableaux for $S_1, S_2$. Let $R_1$ be the set of all $Q_1 \supset Q_1(a_1)$ such that $Q_1(a_1)$ was inferred from $Q_1$ in $\mathscr{I}_1$; let $R_2$ be the set of all $Q_2 \supset Q_2(a_2)$ such that $Q_2(a_2)$ was inferred from $Q_2$ in $\mathscr{I}_2$. Then $\langle R_1, R_2 \rangle$ is a joint associate of $\langle S_1, S_2 \rangle$ (verify!).

We remark that Theorem 3 holds both for unsigned and signed formulas (for the latter we must, of course, in our definition of joint associate everywhere replace "$Q \supset Q(a)$" by "$Q \Rightarrow Q(a)$" and "weak subformula" by "subformula" (cf. § 2 of Chapter XIV).

**A Stronger Form of the Fundamental Theorem.** We now discuss a still stronger form of both the Fundamental Theorem and the Symmetric Form of the Fundamental Theorem.

We wish to simultaneously consider the treatment for both signed and unsigned formulas. Accordingly, if $Q$ is a signed formula, we define $Q \Rightarrow Q(a)$ in the manner of §2 of Chapter XIV; if $Q$ is an unsigned formula, then we shall take $Q \Rightarrow Q(a)$ to be $Q \supset Q(a)$.

By a *Boolean descendant* of a formula $X$ we shall mean a descendant of $X$ in the sense of propositional logic (cf. Chapter III)—i.e. the set of Boolean descendants of $X$ is the smallest set which contains $X$ and which contains with each $\alpha$, both $\alpha_1, \alpha_2$, and with each $\beta$ both $\beta_1, \beta_2$. For unsigned formulas, every Boolean descendant of $X$ is also a weak Boolean subformula of $X$, but not conversely. For signed formulas, every Boolean descendant of $X$ is a subformula of $X$, but not conversely (we call a *signed* formula $\pi_1 X_1$—where $\pi_1$ is "$T$" or "$F$"—a subformula of $\pi_2 X_2$ if $X_1$ is a subformula of $X_2$).

Now we shall call $R$ a *strong* associate of $S$ if $R$ is an associate of $S$, and if in addition, $R$ can be arranged in a regular sequence

$$Q_1 \Rightarrow Q_1(a_1), \dots, Q_n \Rightarrow Q_n(a_n)$$

such that $Q_1$ is a Boolean descendant of some element of $S$, $Q_2$ is a Boolean descendant of some element of $\{S, Q_1(a_1)\}, \dots, Q_n$ is a Boolean descendant of some element of $\{S, Q_1(a_1), \dots, Q_{n-1}(a_{n-1})\}$.

Now suppose $\mathscr{I}$ is a closed analytic tableau for $S$. Let $R$ be the associate of $S$ obtained from $\mathscr{I}$ by the method of §3 of Chapter VII—i.e. $R$ is the set of all formulas $Q \Rightarrow Q(a)$ such that $Q(a)$ was inferred from $Q$ on the tableau $\mathscr{I}$. We assert that $R$ is not only an associate of $S$, but even a strong associate of $S$. We can see this as follows.

Let $Q_1, Q_2, \dots, Q_n$ be the $\gamma$- and $\delta$-formulas which were used in the construction of $\mathscr{I}$ and in the order in which they were used. Let $Q_1(a_1), \dots, Q_n(a_n)$ be the formulas which were respectively inferred from $Q_1, \dots, Q_n$ (thus $R$ can be arranged in the regular sequence $\langle Q_1 \Rightarrow Q_1(a_1), \dots, Q_n \Rightarrow Q_n(a_n) \rangle$). Now before $Q_1$ was used, only truth-functional rules were applied in $\mathscr{I}$, hence $Q_1$ must be a Boolean descendant of some element of $S$. Now let $S_1$ be the set of all formulas which occur on the tableau as of the stage that $Q_1(a_1)$ was first inferred from $Q_1$. No further application of a quantificational rule was made until we inferred $Q_2(a_2)$ from $Q_2$; therefore $Q_2$ must be a Boolean descendant of some element of $S_1$. But every element of $S_1$ except possibly $Q_1(a_1)$ is a Boolean descendant of some element of $S$. And any Boolean descendant of a Boolean descendant of some element of $S$ is again a Boolean descendant of some element of $S$. Therefore $Q_2$ must be a Boolean descendant of some element of $\{S, Q_1(a_1)\}$. Similarly $Q_3$ is a Boolean descendant of some element of $\{S, Q_1(a_1), Q_2(a_2)\}, \ldots, Q_n$ is a Boolean descendant of some element of $\{S, Q_1(a_1), \ldots, Q_{n-1}(a_{n-1})\}$. Thus $R$ is indeed a strong associate of $S$.

An analogous result goes through for the symmetric version. We define $\langle R_1, R_2 \rangle$ to be a *strong joint associate* of $\langle S_1, S_2 \rangle$ if:

(1) $R_1 \cup R_2$ is an associate of $S_1 \cup S_2$.

(2) $R_1$ can be arranged in a regular sequence $\langle Q_1 \Rightarrow Q_1(a_1), \ldots, Q_n \Rightarrow Q_n(a_n) \rangle$ such that for each $i < n$, $Q_{i+1}$ is a Boolean descendant of some element of $\{S_1, Q_1(a_1), \ldots, Q_i(a_i)\}$—and similarly $R_2$ can be arranged in a regular sequence bearing the same relationship to $S_2$.

Now, if $(\mathscr{I}_1, \mathscr{I}_2)$ is a clashing pair for $S_1, S_2$, and if $R_1$ is the set of all $Q \Rightarrow Q(a)$ such that $Q(a)$ was inferred from $Q$ on $\mathscr{I}_1$, and if $R_2$ is the set of all $Q \Rightarrow Q(a)$ such that $Q(a)$ was inferred from $Q$ in $\mathscr{I}_2$, then the reader can verify that $\langle R_1, R_2 \rangle$ is a *strong* joint associate of $\langle S_1, S_2 \rangle$.

We have thus proved

**Theorem 3\*.** (a) *If $S$ is unsatisfiable, then $S$ has a strong associate.*
(b) *If $S_1 \cup S_2$ is unsatisfiable, then $\langle S_1, S_2 \rangle$ has a strong joint associate.*

*Applications.* Consider the Hilbert-type axiom system $Q_2^*$ of Chapter VIII. Suppose we weaken the inference rules by requiring that $Q$ be not only a subformula of $X$, but a *Boolean descendant* of $X$. The resulting system would still be complete by virtue of (a) of Theorem 3\*.

Next we consider the following "symmetric" version of this system which we will term "$QQ$". We shall use Sheffer's stroke symbol (it is not too important whether it is taken as an additional primitive or defined in terms of other logical connectives).

*Axioms:* All tautologies of the form $X \mid Y$.

*Rules: C*:
$$\frac{([\gamma \supset \gamma(a)] \wedge X) \mid Y}{X \mid Y}$$

$$\frac{X \mid ([\gamma \supset \gamma(a)] \wedge Y)}{X \mid Y}$$

provided $\gamma$ is a Boolean descendant of $X$

provided $\gamma$ is a Boolean descendant of $Y$

*D*:
$$\frac{([\delta \supset \delta(a)] \wedge X) \mid Y}{X \mid Y}$$

$$\frac{X \mid ([\delta \supset \delta(a)] \wedge Y)}{X \mid Y}$$

provided $\delta$ is a Boolean descendant of $X$ and $a$ does not occur in $X \mid Y$

provided $\delta$ is a Boolean descendant of $Y$ and $a$ does not occur in $X \mid Y$

The system $QQ$ is complete in the sense that for any two sentences $X$, $Y$ which are incompatible, $X|Y$ is provable in $QQ$. This is proved using (b) of Theorem 3* in virtually the same manner as we proved the completeness of the system $Q_2^*$, using the Fundamental Theorem.

The system $QQ$ could have been alternatively presented as a GENTZEN type system of stroke sequents as follows (the system works if $U$, $V$ are construed as sets of signed or unsigned formulas).

*Axioms:* All tautologies $U|V$.

*Rules: C:*

$$\frac{U, \gamma \supset \gamma(a)|V}{U|V} \qquad\qquad \frac{U|V, \gamma \supset \gamma(a)}{U|V}$$

provided $\gamma$ is a Boolean descendant of some element of $U$

provided $\gamma$ is a Boolean descendant of some element of $V$

*D:*

$$\frac{U, \delta \supset \delta(a)|V}{U|V} \qquad\qquad \frac{U|V, \delta \supset \delta(a)}{U|V}$$

provided $\delta$ is a Boolean descendant of some element of $U$ and $a$ does not occur in $U|V$

provided $\delta$ is a Boolean descendant of some element of $V$ and $a$ does not occur in $U|V$.

The system $QQ$ in the above Gentzen-type formulation is complete in the sense that every valid stroke sequent is provable.

The completeness of the system $QQ$ affords another proof of Craig's Interpolation lemma, reducing it to the case for Propositional Logic. (Craig's lemma for propositional logic can be proved in a different method—perhaps more simple—than that of the preceding chapter; we shall do this in the next chapter.) First of all, by an interpolation formula for a *stroke* sequent $U|V$ we shall mean a formula $X$ such that all predicates and parameters of $X$ occur both in $U$ and in $V$, and $U \rightarrow X$, $V \rightarrow \bar{X}$ are both valid. Alternatively, an interpolation formula for a stroke sequent $U|V$ is the same as an interpolation formula for the Gentzen sequent $U \rightarrow \bar{V}$ (because the validity of $X \rightarrow \bar{V}$ is equivalent to the validity of $V \rightarrow \bar{X}$). Thus the existence of interpolation formulas for all valid stroke-sequents is equivalent to the existence of interpolation formulas for all valid Gentzen sequents.

Now, assume the Interpolation Lemma for propositional logic. To prove the Interpolation Lemma for First Order Logic, it suffices (by virtue of the completeness of $QQ$) to show that there is an interpolation formula for each of the axioms, and in any application of an inference rule, if there is an interpolation formula for the premise, then there is an interpolation formula for the conclusion. Since we are taking all tautologies as axioms, then the existence of interpolation formulas for all axioms is simply the interpolation lemma for propositional logic. Now for the inference rules:

$C$: Suppose $X$ is an interpolation formula for $U, \gamma \supset \gamma(a) | V$, where $\gamma$ is a Boolean descendant of (some element of) $U$. We assert that if $a$ does not occur in $X$, or if $a$ does occur in some element of $U$, then $X$ is an interpolation formula for $U | V$; if $a$ does not occur in $U$ but does occur in $X$, $(\forall x) X_x^a$ is an interpolation formula for $U | V$ (where $x$ is any variable which has no free occurrence in $X$). To verify this, the key point is that since $\gamma$ is a *Boolean* descendant of some element of $U$, then not only does every predicate of $\gamma$ occur in $U$ (which would be the case even if $\gamma$ were only a descendant of $U$) but also every *parameter* of $\gamma$ occurs in $U$. The detailed verification is as follows. Since $U, \gamma \supset \gamma(a) \rightarrow X$ is valid, so is $U \rightarrow X$ (because $\gamma \supset \gamma(a)$ is valid). And we are given that $V \rightarrow \sim X$ is valid. Suppose that $a$ does not occur in $X$ or that $a$ occurs in $U$. Every predicate $P$ of $X$ occurs in $U$ or in $\gamma \supset \gamma(a)$; if $P$ occurs in $\gamma \supset \gamma(a)$ then $P$ occurs in $\gamma$, and hence in $U$. So all predicates of $X$ occur in $U$. As for the parameters, we are assuming that if $a$ is in $X$, it is also in $U$. Now consider any parameter $b$ of $X$ distinct from $a$. We know that $b$ is in $U$ or in $\gamma \supset \gamma(a)$. Suppose $b$ is in $\gamma \supset \gamma(a)$. Then $b$ must be in $\gamma$ or in $\gamma(a)$. Since $a$ is the only parameter which could be in $\gamma(a)$ but not in $\gamma$, then $b \in \gamma$. But all parameters of $\gamma$ are in $U$. Thus $b \in U$. So all parameters of $X$ are in $U$, hence $X$ is indeed an interpolation formula for $U | V$.

Now consider the case that $a$ is in $X$ but not in $U$. Since $U \rightarrow X$ is valid and $a$ does not occur in $U$, then $U \rightarrow (\forall x) X_x^a$ is valid. Since $V \rightarrow \sim X$ is valid, then $V \rightarrow \sim (\forall x) X_x^a$ is valid. Every predicate of $(\forall x) X_x^a$ is also a predicate of $X$, and we already know that all predicates of $X$ are in both $U$ and $V$. As for the parameters, let $b$ be any parameter of $(\forall x) X_x^a$. Then $b \neq a$ (because $a$ does not occur in $(\forall x) X_x^a$. Since $b$ occurs in $(\forall x) X_x^a$, then $b$ occurs in $X$. Hence $b$ occurs in $\{U, \gamma \supset \gamma(a)\}$. Then $b$ must occur in $U$ for the same reason as before. Also $b$ occurs in $V$, since it occurs in $X$. Thus we have all the conditions of $(\forall x) X_x^a$ being an interpolation formula for $U | V$.

We have thus verified the left half of Rule $C$. The right half then easily results by the following argument. Suppose there is an interpolation formula $X$ for $U | V, \gamma \supset \gamma(a)$ and that $\gamma$ is a Boolean descendant of $V$. Then $\sim X$ is an interpolation formula for $V, \gamma \supset \gamma(a) | U$. Hence by our previous argument for the left half of Rule $C$, there is an interpolation formula $Y$ for $V | U$ ($Y$ is either $\sim X$ or $(\forall x)(\sim X)_x^a$). Then $\sim Y$ is an interpolation formula for $U | V$.

$D$: Suppose $X$ is an interpolation formula for $U, \delta \supset \delta(a) | V$ and that $\delta$ is a Boolean descendant of $U$, and that $a$ does not occur in $U$ nor $V$.

Then $X$ is an interpolation formula for $U|V$. Reason: Since all parameters of $X$ occur in $V$ and $a$ does not occur in $V$, then $a$ does not occur in $X$. Now, $U, \delta \supset \delta(a) \to X$ is valid and $a$ does not occur in $U$ nor $X$, hence $U \to X$ is valid. All predicates of $X$ are in $U$ by the same argument as for Rule $C$. Let $b$ be any parameter of $X$. Then $b \neq a$ (since $b$ occurs in $V$ and $a$ doesn't), and $b$ must occur in $U$ by the same argument as for Rule $C$. Also, of course, $V \to \sim X$ is valid and all predicates and parameters of $X$ occur in $V$. So $X$ is an interpolation formula for $U \mid V$. This proves the left half of Rule $D$. The right half is then derivable from the left half in a manner similar to the argument for Rule $C$.

Chapter XVII

# Systems of Linear Reasoning

In Craig's paper "Linear Reasoning" [1] he considered a system of Quantification Theory in which there are no axioms, but only inference rules, and all rules are 1-premise rules. In each of these rules, the premise validly implies the conclusion. By a *derivation* of $Y$ from $X$ in the system is meant a finite sequence of lines

$$X_1 \quad (= X)$$
$$\vdots$$
$$X_n \quad (= Y)$$

such that for each $i < n$, the line $X_{i+1}$ is a direct consequence of the preceding line $X_i$ by one of the inference rules.

His system is complete in the sense that for any conjunction $X$ of prenex sentences and any disjunction $Y$ of prenex sentences, if $X \supset Y$ is valid, then there exists a derivation of $Y$ from $X$—moreover one in which some intermediate line $X_i$ is an interpolation formula for $X \supset Y$! (This constituted the first known proof of Craig's interpolation lemma. It was subsequent authors who extricated the proof of the interpolation lemma from the completeness theorem for Craig's system of linear reasoning, the latter being a far more complicated matter.)

In this chapter, we will not study Craig's original system, but will rather present other systems of linear reasoning which arise very naturally from the tableau point of view. The system which we first consider—which we tend to regard as the "basic" system—does not require appeal to prenex normal form. We obtain yet another approach to propositional logic and quantification theory, somewhat like tableaux, but without use of trees. The completeness proof for the system is based primarily on the completeness theorem for clashing tableaux which we proved in the last chapter. Then we consider some alternative systems based on our previous results on $P$-clashing tableaux and the strong symmetric form of the fundamental theorem.

## § 1. Configurations

By a *configuration* $\mathscr{C}$ we shall mean a finite collection of finite sets of sentences. If $\mathscr{C} = \{B_1, \ldots, B_n\}$ we also write $\mathscr{C}$ in the form

$$\boxed{B_1} \quad \boxed{B_2} \cdots \boxed{B_n}$$

and we refer to the elements $B_1, \ldots, B_n$ of $\mathscr{C}$ as the *blocks* of the configuration. We shall sometimes write $\boxed{B}$ to mean the configuration whose only element is the set $B$—such a configuration is a 1-block configuration.

Informally we read $\mathscr{C}$ as saying "either all the elements of $B_1$ are true or all the elements of $B_2$ are true or ... or all the elements of $B_n$ are true." More precisely, we define $\mathscr{C}$ to be *true* under an interpretation (or valuation) $I$ if $\mathscr{C}$ contains at least one block all of whose elements are true under $I$. (In particular, the empty configuration is always *false*, whereas the configuration whose only block is the empty set is always *true*.) Then we define a configuration to be *valid* if it is true in all interpretations, and *satisfiable* if it is true in at least one. We define two configurations $\mathscr{C}_1$, $\mathscr{C}_2$ to be *equivalent* if they are true in the same interpretations. We say that $\mathscr{C}_1$ *implies* $\mathscr{C}_2$ if $\mathscr{C}_2$ is true in every interpretation in which $\mathscr{C}_1$ is true. Let us note that if we remove one or more elements from any block of a configuration $\mathscr{C}_1$, the resulting configuration $\mathscr{C}_2$ is implied—indeed *truth-functionally* implied—by $\mathscr{C}_1$ (i.e. $\mathscr{C}_2$ is true in all Boolean valuations which satisfy $\mathscr{C}_1$).

Let $\mathscr{C}$ be a configuration $\boxed{B_1} \cdots \boxed{B_n}$ in which the blocks consists of *unsigned* formulas. Then the configuration $\mathscr{C}$ is equivalent to the single formula $\hat{B}_1 \vee \cdots \vee \hat{B}_n$ (if $B_i$ is empty, then by $\hat{B}_i$ we shall mean $t$). We shall refer to $\hat{B}_1 \vee \cdots \vee \hat{B}_n$ as a *translation* of the configuration $\mathscr{C}$. If $\mathscr{C}$ is empty, then we define its translation to be $f$.

For any finite set $S$ of *signed* formulas, by $\hat{S}$ we shall mean the result of first *unsigning* the formulas (i.e. deleting "$T$"s and replacing "$F$"s by "$\sim$"s) and then taking the conjunction (the order is immaterial). Then for a configuration $\boxed{B_1} \cdots \boxed{B_n}$ whose blocks $B_i$ consist of *signed* formulas, we again call the formula $\hat{B}_1 \vee \cdots \vee \hat{B}_n$ a *translation* of the configuration

$\boxed{B_1}\cdots\boxed{B_n}$ We remark that the translation of a configuration is not unique, but any two translations are equivalent—indeed truth-functionally equivalent—and contain the same predicates and parameters.

Now we define $\mathscr{C}_2$ to be a *reduction* of $\mathscr{C}_1$ if $\mathscr{C}_2$ can be obtained from $\mathscr{C}_1$ by finitely many of the following operations:

$A$: Replace any block $\boxed{S,\alpha}$     by     $\boxed{S,\alpha_1,\alpha_2}$ ;

$B$: Replace any block $\boxed{S,\beta}$     by     $\boxed{S,\beta_1}$   $\boxed{S,\beta_2}$ (i.e. we

     replace $\boxed{S,\beta}$    by the *two* blocks    $\boxed{S,\beta_1}$ ,   $\boxed{S,\beta_2}$ ;

$C$: Replace any block $\boxed{S,\gamma}$     by     $\boxed{S,\gamma,\gamma(a)}$ ;

$D$: Replace any block $\boxed{S,\delta}$     by     $\boxed{S,\delta(a)}$ , provided that $a$ is *new* to the configuration.

If $\mathscr{C}$ is a configuration $\boxed{S_1}\cdots\boxed{S_n}$, then by $\mathscr{C}\ \boxed{S}$ we mean $\boxed{S_1}\cdots\boxed{S_n}\ \boxed{S}$; by $\mathscr{C}\ \boxed{S}\ \boxed{S'}$ we mean $\boxed{S_1}\cdots\boxed{S_n}\ \boxed{S}\ \boxed{S'}$, etc. We thus can display our reduction operation in the following schematic form:

$A$:    $\dfrac{\mathscr{C}\ \boxed{S,\alpha}}{\mathscr{C}\ \boxed{S,\alpha_1,\alpha_2}}$

$B$:    $\dfrac{\mathscr{C}\ \boxed{S,\beta}}{\mathscr{C}\ \boxed{S,\beta_1}\ \boxed{S,\beta_2}}$

$C$:    $\dfrac{\mathscr{C}\ \boxed{S,\gamma}}{\mathscr{C}\ \boxed{S,\gamma,\gamma(a)}}$

$D$:    $\dfrac{\mathscr{C}\ \boxed{S,\delta}}{\mathscr{C}\ \boxed{S,\delta(a)}}$,   provided $a$ is new to $\mathscr{C}\ \boxed{S,\delta}$.

We call a block *closed* if it contains some element $X$ and its conjugate $\bar{X}$. If the block contains some *atomic* $X$ and $\bar{X}$, then we call the block *atomically* closed. We call a configuration closed if each of its blocks is closed, and atomically closed if each of its blocks is atomically closed. A closed configuration is obviously unsatisfiable.

We leave it to the reader to verify that any reduction of a satisfiable configuration is again satisfiable. Therefore if $\mathscr{C}$ is reducible to a closed configuration, then $\mathscr{C}$ is unsatisfiable. In particular, if a 1-block configuration $\boxed{S}$ is reducible to a closed configuration, then $S$ must be unsatisfiable.

Next we wish to show that if $S$ is unsatisfiable, then $\boxed{S}$ is reducible to an *atomically* closed configuration. This "completeness" theorem is an easy consequence of the completeness theorem for modified block tableaux. For any modified block tableau $\mathscr{T}$, define its *corresponding* configuration $\mathscr{C}$ to be the set of end points of $\mathscr{T}$. Now, it is obvious that if $\mathscr{C}$ is the corresponding configuration of $\mathscr{T}$, then $\mathscr{C}$ is a reduction of the origin of $\mathscr{T}$. And if $\mathscr{T}$ is atomically closed, so is $\mathscr{C}$. Now, if $S$ is unsatisfiable, then there is an atomically closed tableau for $S$. Hence the corresponding configuration $\mathscr{C}$ of $\mathscr{T}$ is an atomically closed reduction of $S$. This proves:

**Theorem 1.** *S is unsatisfiable iff* $\boxed{S}$ *is reducible to an atomically closed configuration.*

*Exercise 1.* Define $S$ to be inconsistent if $\boxed{S}$ is reducible to an atomically closed configuration. Establish Theorem 1 as a consequence of the Unifying Principle.

*Exercise 2.* Show that a configuration $\mathscr{C}$ (with possibly more than one block) is unsatisfiable iff $\mathscr{C}$ is reducible to a closed configuration.

*Clashing Configurations.* We shall say that an ordered pair $\langle \mathscr{C}_1, \mathscr{C}_1' \rangle$ of configurations is *reducible* to an ordered pair $\langle \mathscr{C}_n, \mathscr{C}_n' \rangle$ if there exists a finite sequence $\langle \mathscr{C}_1, \mathscr{C}_1' \rangle, \ldots, \langle \mathscr{C}_n, \mathscr{C}_n' \rangle$—which we call a *reduction sequence*— such that for each $i < n$, either $\mathscr{C}_{i+1}$ is obtained from $\mathscr{C}_i$ or $\mathscr{C}_{i+1}'$ is obtained from $\mathscr{C}_i'$ by (one application of) one of the operations $A$, $B$, $C$, $D$ and moreover if it is operation $D$, then the parameter $a$ must be new to *both* $\mathscr{C}_i$ and $\mathscr{C}_i'$. If $\langle \mathscr{C}_1, \mathscr{C}_1' \rangle$ is reducible to $\langle \mathscr{C}_2, \mathscr{C}_2' \rangle$, and if the former pair is jointly satisfiable (i.e. if there is an interpretation in which $\mathscr{C}_1, \mathscr{C}_1'$ are both true) then the latter pair is also jointly satisfiable (why?). We shall say that a pair $\langle \mathscr{C}, \mathscr{C}' \rangle$ of configurations *clash* if for each block $B$ of $\mathscr{C}$ and each block $B'$ of $\mathscr{C}'$, $B \cup B'$ is *atomically* closed. Obviously a clashing pair of configurations is not jointly satisfiable, hence any pair of configurations which is reducible to a clashing pair is not jointly satisfiable.

If $\langle \mathscr{I}_1, \mathscr{I}_2 \rangle$ is a joint pair of modified block tableaux for $S_1 | S_2$, and if $\mathscr{C}_1$, $\mathscr{C}_2$ are the configurations corresponding to $\mathscr{I}_1$, $\mathscr{I}_2$ respectively, then $\langle \mathscr{C}_1, \mathscr{C}_2 \rangle$ is a reduction of $\langle \boxed{S_1}, \boxed{S_2} \rangle$ (verify!). We know from § 1 of the last chapter that if $S_1$ is incompatible with $S_2$, then there does exist a clashing pair $\langle \mathscr{T}_1, \mathscr{T}_2 \rangle$ of modified block tableaux for $S_1 | S_2$. The corresponding configurations $\mathscr{C}_1, \mathscr{C}_2$ obviously clash. We have thus shown:

**Theorem 2.** $S_1$ *is incompatible with* $S_2$ *iff* $\langle \boxed{S_1}, \boxed{S_2} \rangle$ *is reducible to a clashing pair of configurations.*

**Remarks.** We wish to discuss how a clashing pair $\langle \mathscr{C}_1, \mathscr{C}_2 \rangle$ which is a reduction of $\langle \boxed{S_1}, \boxed{S_2} \rangle$ yields an interpolation formula for the stroke sequent $S_1 | S_2$. In the case of *propositional* logic, the matter is quite simple. Of course in propositional logic, our *reduction* procedure uses only operations $A$ and $B$, and clearly if $\mathscr{C}_1$ is reducible to $\mathscr{C}_2$ using only operations $A$ and $B$, then the configuration $\mathscr{C}_1$ is *equivalent* to the configuration $\mathscr{C}_2$. We might also make the following remark: In both propositional logic and quantification theory, if $\langle \mathscr{C}'_1, \mathscr{C}'_2 \rangle$ is a reduction of $\langle \mathscr{C}_1, \mathscr{C}_2 \rangle$, then certainly $\mathscr{C}'_1$ is a reduction of $\mathscr{C}_1$ and $\mathscr{C}'_2$ is a reduction of $\mathscr{C}_2$. Now suppose conversely that $\mathscr{C}'_1$ is a reduction of $\mathscr{C}_1$ and $\mathscr{C}'_2$ is a reduction of $\mathscr{C}_2$. Then in quantification theory, $\langle \mathscr{C}'_1, \mathscr{C}'_2 \rangle$ is generally *not* a reduction of $\langle \mathscr{C}_1, \mathscr{C}_2 \rangle$ (because of operation $D$), but in the case of propositional logic, it is a reduction.

Now let $S_1, S_2$ be (finite) sets of signed formulas of propositional logic, and let $\langle \mathscr{C}_1, \mathscr{C}_2 \rangle$ be a reduction of $\langle \boxed{S_1}, \boxed{S_2} \rangle$ which *clashes*. To find an interpolation formula for $S_1 | S_2$ we proceed as follows. First we delete all closed blocks from $\mathscr{C}_1$; the resulting configuration $\mathscr{C}'_1$ is obviously equivalent to $\mathscr{C}_1$. Then we delete all *non-atomic* elements from all blocks of $\mathscr{C}'_1$, and also all *atomic* elements $X$ such that $\bar{X}$ does not occur in any block of $\mathscr{C}_2$. Then the resulting configuration—call it $\mathscr{C}_1 \upharpoonright \mathscr{C}_2$—still clashes with $\mathscr{C}_2$ (why?), and is *implied* by $\mathscr{C}_1$, and all propositional variables which occur in any block $B_i$ of $\mathscr{C}_1 \upharpoonright \mathscr{C}_2$ must occur in $\mathscr{C}_2$, hence they occur in *both* $S_1$ and in $S_2$. Now let $X$ be a translation of $\mathscr{C}_1 \upharpoonright \mathscr{C}_2$ and consider the signed formula $TX$. Clearly $S_1 \to TX$ holds. And since $\mathscr{C}_1 \upharpoonright \mathscr{C}_2$ clashes with $\mathscr{C}_2$, and $\mathscr{C}_2$ is equivalent to $\boxed{S_2}$ and $TX$ is equivalent to $\mathscr{C}_1 \upharpoonright \mathscr{C}_2$, then $TX$ is incompatible with $S_2$—i.e. $S_2 \to FX$ holds. And all propositional variables of $X$ occur in both $S_1$ and $S_2$. So $TX$ is an interpolation formula for $S_1 | S_2$.

**Further Remarks.** At the propositional level, the use of configurations to obtain interpolation formulas is tantamount to reduction to disjunctive normal form (cf. Exercise 3, end of Chapter I). Indeed, suppose $X$ is truth-functionally incompatible with $Y$. Let $C_1 \vee \cdots \vee C_n$ be a reduction of $X$ to disjunctive normal form. If in each $C_i$ we delete all conjunctive components that do not occur in $Y$, the resulting disjunctive normal formula $C_1' \vee \cdots \vee C_n'$ is still incompatible with $Y$, and is still implied by $X$, and is hence an interpolation formula for $X \supset (\sim Y)$.

**Quantified Configurations.** For Quantification Theory, the use of clashing configurations to obtain interpolation formulas is more complicated. The trouble is this: In propositional logic, our "reduction" operations $A$, $B$ are such that any reduction of a configuration $\mathscr{C}$ is equivalent to $\mathscr{C}$, but (as we have already remarked), this is not the case in quantification theory because of operation $D$. We now remedy the situation as follows.

By a *parameter prefix*—which we will also refer to just as a "prefix"— we shall mean an expression $(q_1 a_1)(q_2 a_2) \cdots (q_n a_n)$, which we also write as $q_1 a_1 q_2 a_2 \ldots q_n a_n$, where $a_1, \ldots, a_n$ are *distinct parameters* and each $q_i$ is one of the symbols "$\forall$", "$\exists$" (e.g. $\forall a_1 \exists a_2 \forall a_3 \forall a_4$ is a parameter prefix). We also allow the empty prefix. Now consider a prefix $q_1 a_1 \ldots q_n a_n$ and an unsigned formula $X$ and consider the expression $q_1 a_1 \ldots q_n a_n X$. Strictly speaking, this expression is not a formula, since our formation rules do not allow quantifying parameters. But for our present purposes, it will be convenient to "pretend" that $a_1, \ldots, a_n$ are variables rather than parameters. To be precise, let $x_1, \ldots, x_n$ be variables which do not occur in $X$ (and such that $x_i \neq x_j$ for $i \neq j$); let $X'$ be the result of substituting $x_1$ for $a_1, \ldots, x_n$ for $a_n$ in $X$. Call $q_1 x_1 \ldots q_n x_n X'$ a *translation* of $q_1 a_1 \ldots q_n a_n X$. By an interpretation $I$ of the expression $q_1 a_1 \ldots q_n a_n X$ we mean an assignment of values to all predicates of $X$ and to all parameters of $X$ *other than* $a_1, \ldots, a_n$. Now we define the expression $q_1 a_1 \ldots q_n a_n X$ to be *true* under $I$ if a translation $q_1 x_1 \ldots q_n x_n X'$ is true under $I$.

By a quantified configuration we shall mean an ordered pair $\langle \sigma, \mathscr{C} \rangle$ —also written $\sigma \mathscr{C}$—where $\mathscr{C}$ is a configuration and $\sigma$ is a prefix such that every parameter which occurs in $\sigma$ also occurs in $\mathscr{C}$ (i.e. in at least one element of at least one block of $\mathscr{C}$). (It will prove technically convenient to arrange matters so that we never have "vacuous" quantifiers in our prefix.) By a *translation* of $\sigma \mathscr{C}$ we shall mean a translation of $\sigma X$, where $X$ is a translation of $\mathscr{C}$. By an *interpretation $I$* of $\sigma \mathscr{C}$ we mean an interpretation of $\sigma X$, where $X$ is a translation of $\mathscr{C}$, and we call $\sigma \mathscr{C}$ *true* under $I$ if $\sigma X$ is true under $I$.

Preparatory to our consideration of "reduction" operations for quantified configurations, we make the following observations. Consider two prefixes $\sigma_1, \sigma_2$ which contain exactly the same parameters. For such a pair we shall say that $\sigma_1$ *implies* $\sigma_2$, or that $\sigma_2$ is *weaker* than $\sigma_1$, if for any formula $X$ with the same parameters as $\sigma_1$ (or $\sigma_2$), $\sigma_1 X$ implies $\sigma_2 X$. Let us note the following implications of prefixes:

(1)  $\forall a \forall b$  implies  $\forall b \forall a$.

(2)  $\exists a \forall b$  implies  $\forall b \exists a$  (but not conversely).

(3)  $\exists a \exists b$  implies  $\exists b \exists a$.

(4)  $\forall a$  implies  $\exists a$.

More generally:

(1)  $\sigma_1 \forall a \forall b \sigma_2$  implies  $\sigma_1 \forall b \forall a \sigma_2$.

(2)  $\sigma_1 \exists a \forall b \sigma_2$  implies  $\sigma_1 \forall b \exists a \sigma_2$.

(3)  $\sigma_1 \exists a \exists b \sigma_2$  implies  $\sigma_1 \exists b \exists a \sigma_2$.

(4)  $\sigma_1 \forall a \sigma_2$  implies  $\sigma_1 \exists a \sigma_2$.

This means that in a prefix, if we move a universal quantifier any number of places to the left, or move an existential quantifier any number of places to the right, or if we change a universal quantifier to an existential quantifier, we weaken the prefix. Thus $\sigma_1 \sigma_2 \forall a$ implies both $\sigma_1 \forall a \sigma_2$ and $\sigma_1 \exists a \sigma_2$. Using our notation "$q$" to mean either "$\forall$" or "$\exists$", we can say $\sigma_1 \sigma_2 \forall a$ implies $\sigma_1 q a \sigma_2$, regardless of whether $q$ is "$\forall$" or "$\exists$".

The following are our reduction operations for quantified configurations:

$$A: \quad \frac{\sigma \mathscr{C} \;\boxed{S, \alpha}}{\sigma \mathscr{C} \;\boxed{S, \alpha_1, \alpha_2}}$$

$$B: \quad \frac{\sigma \mathscr{C} \;\boxed{S, \beta}}{\sigma \mathscr{C} \;\boxed{S, \beta_1} \;\boxed{S, \beta_2}}$$

$C: \quad \dfrac{\sigma \mathscr{C} \;\boxed{S, \gamma}}{\sigma' \mathscr{C} \;\boxed{S, \gamma, \gamma(a)}}$, where $\sigma'$ is defined as follows: If $a$ is in $\sigma$ or in $\mathscr{C} \;\boxed{S, \gamma}$ or not in $\mathscr{C} \;\boxed{S, \gamma, \gamma(a)}$, then $\sigma' = \sigma$. Otherwise, $\sigma'$ is any prefix obtained by inserting either $\forall a$ or $\exists a$ anywhere in $\sigma$.

$D: \quad \dfrac{\sigma \mathscr{C} \;\boxed{S, \delta}}{\sigma' \mathscr{C} \;\boxed{S, \delta(a)}}$, where $a$ does not occur in $\delta$ nor in $\mathscr{C} \;\boxed{S, \delta}$ and $\sigma'$ is defined as follows: If $a$ does not occur in $\delta(a)$, then $\sigma' = \sigma$. Otherwise $\sigma' = \sigma \exists a$.

We remark that in operations $A, B, D$ the conclusion is *equivalent* to the premise. As for operation $C$, if the first case holds (i.e. if $a$ is in $\sigma$ or in $\mathscr{C} \;\boxed{S, \gamma}$ or not in $\mathscr{C} \;\boxed{S, \gamma, \gamma(a)}$), then the conclusion is equivalent to the premise. In the second case, then $\sigma \mathscr{C} \boxed{S, \gamma}$ is equivalent to $\sigma \forall a$ $\mathscr{C} \boxed{S, \gamma, \gamma(a)}$, which in turn implies $\sigma' \mathscr{C} \boxed{S, \gamma, \gamma(a)}$. Thus in all four operations, the premise always implies the conclusion.

An unquantified configuration $\mathscr{C}$ is a logically distinct object from the quantified configuration $\Phi \mathscr{C}$ (where $\Phi$ is the empty prefix) even though they are true under precisely the same interpretations. In particular, by a reduction of $\mathscr{C}$ we mean a reduction using the original operations $A, B, C, D$; by a reduction of $\Phi \mathscr{C}$ is meant a reduction using the new operations $A, B, C, D$. Thus a reduction of $\mathscr{C}$ is not in general implied by $\mathscr{C}$, whereas a reduction of $\Phi \mathscr{C}$ is implied by $\mathscr{C}$.

We similarly define a reduction of an ordered pair $\langle \sigma_1 \mathscr{C}_1, \sigma_2 \mathscr{C}_2 \rangle$ (but in using operation $D$, the parameter $a$ must be new to both quantified configurations). More precisely, to perform a reduction operation on an ordered pair $\langle \sigma_1 \mathscr{C}_1, \sigma_2 \mathscr{C}_2 \rangle$ we mean the act of performing the operation on just the first or second member of the pair, doing nothing to the other member, and if the operation is $D$, we demand that the principal parameter $a$ must be new to both $\mathscr{C}_1$ and $\mathscr{C}_2$.

For any prefixes $\sigma_1$, $\sigma_2$, by $\sigma_1 \upharpoonright \sigma_2$ we mean the result of deleting from $\sigma_1$ all quantifiers $qa$ such that $a$ does not occur in $\sigma_2$. (E.g. if $\sigma_1 = \forall a \, \exists b \, \exists c \, \forall d$ and $\sigma_2 = \exists e \, \forall a \, \exists f \, \forall c$, then $\sigma_1 \upharpoonright \sigma_2 = \forall a \, \exists c$, and $\sigma_2 \upharpoonright \sigma_1 = \forall a \, \forall c$.) By the *dual* $\sigma^0$ of a prefix $\sigma$ is meant the result of changing each "$\forall$" to "$\exists$" and each "$\exists$" to "$\forall$" (e.g. the dual of $\forall a \, \exists b \, \forall c$ is $\exists a \, \forall b \, \exists c$). Obviously $\sigma_1 \upharpoonright \sigma_2^0$ is the same as $\sigma_1 \upharpoonright \sigma_2$. We shall call a pair $(\sigma_1, \sigma_2)$ of prefixes a *special* pair if $\sigma_1 \upharpoonright \sigma_2$ is the dual of $\sigma_2 \upharpoonright \sigma_1$—equivalently, if $\sigma_1 \upharpoonright \sigma_2^0 = (\sigma_2 \upharpoonright \sigma_1)^0$. The following lemma is basic.

**Lemma I.** *If* $\langle \mathscr{C}_1, \mathscr{C}_2 \rangle$ *is reducible to* $\langle \mathscr{D}_1, \mathscr{D}_2 \rangle$ *and if* $(\sigma_1, \sigma_2)$ *is special, then there is a special pair* $(\psi_1, \psi_2)$ *such that* $\langle \sigma_1 \mathscr{C}_1, \sigma_2 \mathscr{C}_2 \rangle$ *is reducible to* $\langle \psi_1 \mathscr{D}_1, \psi_2 \mathscr{D}_2 \rangle$.

**Proof.** By an obvious induction argument, it suffices to prove the lemma for the case that $\langle \mathscr{D}_1, \mathscr{D}_2 \rangle$ is obtained from $\langle \mathscr{C}_1, \mathscr{C}_2 \rangle$ by just one reduction step. If the step is operation $A$ or $B$, then we obviously take $\psi_1$ to be $\sigma_1$ and $\psi_2$ to be $\sigma_2$. Now suppose the step is an application of operation $C$— applied say to the left member $\mathscr{C}_1$ (a similar argument would apply if the application were to the right member $\mathscr{C}_2$). Let $a$ be the principal parameter of the application. Of course we take $\psi_2$ to be $\sigma_2$. As for $\psi_1$, if the first case holds, then we have no choice but to take $\psi_1$ to be $\sigma_1$, in which case certainly $(\psi_1, \psi_2)$ is special. Suppose the second case holds (i.e. $a$ is not in $\sigma_1$ nor $\mathscr{C}_1$ but is in $\gamma(a)$). If $a$ does not occur in $\sigma_2$, then $(\sigma_1 \forall a, \sigma_2)$ is again a special pair, so we then take $\psi_1$ to be $\sigma_1 \forall a$. If $a$ is in $\sigma_2$, then we take $q$ to be "$\exists$" if $\forall a$ occurs in $\sigma_2$ or "$\forall$" if $\exists a$ occurs in $\sigma_2$; then we must put $qa$ somewhere in the prefix $\sigma_1$ so that the resulting prefix $\psi_1$ will be such that $(\psi_1, \sigma_2)$ is again special. We do this as follows. Let $q_1 b$ be the rightmost quantifier of $\sigma_1$ such that $b$ is to the left of $a$ in $\sigma_2$ and let $q_2 c$ be the leftmost quantifier of $\sigma_1$ such that $c$ occurs to the right of $a$ in $\sigma_2$. Then we can insert $qa$ anywhere in $\sigma_1$ between $q_1 b$ and $q_2 c$. This takes care of operation $C$. As to operation $D$, there is little to prove! We must take $\psi_2 = \sigma_2$ and $\psi_1$ to be $\sigma_1$ if $a$ does not occur in $\delta(a)$, or $\sigma \exists a$ otherwise. In the first case, $(\psi_1, \psi_2) = (\sigma_1, \sigma_2)$ hence is special. In the second case, since $a$ does not occur in $\sigma_2$ (because it does not occur in $\mathscr{C}_2$), then $(\sigma_1 \exists a, \sigma_2)$ is again special. This concludes the proof.

Next we consider, in addition to the reduction operations $A, B, C, D$, the following two operations:

$E$:  (Existential Generalization)

$$\frac{\sigma \mathscr{C}}{\sigma \exists a \mathscr{C}}, \text{ provided } a \text{ occurs in } \mathscr{C} \text{ but not in } \sigma.$$

$F$:  (Deletion)

$\dfrac{\sigma \mathscr{C}}{\sigma' \mathscr{C}'}$,  where $\mathscr{C}'$ is the result of either removing a closed block from $\mathscr{C}$ or removing an element from an open block of $\mathscr{C}$, and $\sigma'$ is the result of deleting from $\sigma$ all quantifiers $qa$ such that $a$ does not occur in $\mathscr{C}'$.

In operations $E$ and $F$, the premise implies the conclusion (verify!). To perform operation $E$ or $F$ on an ordered pair $\langle \sigma_1 \mathscr{C}_1, \sigma_2 \mathscr{C}_2 \rangle$ of configurations, we mean the act of performing $E, F$ on either $\sigma_1 \mathscr{C}_1$ or $\sigma_2 \mathscr{C}_2$.

The main result we wish to show is that if $S_1$ is incompatible with $S_2$, then by using operations $A, B, C, D, E, F$ it is possible to reduce $\langle \Phi \boxed{S_1}$,

$\Phi \boxed{S_2} \rangle$ to a pair $\langle \sigma_1 \mathscr{C}_1, \sigma_2 \mathscr{C}_2 \rangle$ which satisfies the following four conditions:

(1) $\mathscr{C}_1$ clashes with $\mathscr{C}_2$.
(2) All blocks of $\mathscr{C}_1$ and of $\mathscr{C}_2$ contain only atomic elements.
(3) An element occurs in some block of $\mathscr{C}_1$ iff its conjugate occurs in $\mathscr{C}_2$.
(4) $\sigma_2$ is the dual of $\sigma_1$.

A pair $\langle \sigma_1 \mathscr{C}_1, \sigma_2 \mathscr{C}_2 \rangle$ satisfying conditions (1), (2), (3), (4) above we shall call a *basic* pair. Such a pair is incompatible (i.e. not jointly satisfiable) (why?). We re-state the result we wish to prove as

**Theorem 3.** *If $S_1$ is incompatible with $S_2$, then* $\langle \Phi \boxed{S_1}, \Phi \boxed{S_2} \rangle$ *is $A - F$ reducible (i.e. reducible using operations $A, B, C, D, E, F$) to a basic pair* $\langle \sigma_1 \mathscr{C}_1, \sigma_2 \mathscr{C}_2 \rangle$.

**Proof.** Suppose $S_1$ is incompatible with $S_2$. Then by Theorem 2, $\langle\, \boxed{S_1}\,,\, \boxed{S_2}\,\rangle$ is reducible to a clashing pair $\langle \mathscr{D}_1, \mathscr{D}_2 \rangle$. Let $a_1, \ldots, a_n$ be the parameters (in any order) of $S_1$ which do not occur in $S_2$; let $b_1, \ldots, b_j$ be the parameters common to $S_1, S_2$, and let $c_1, \ldots, c_t$ be the parameters of $S_2$ which do not occur in $S_1$. Let $\psi_1$ be the prefix $\exists\, a_1, \ldots, \exists\, a_n$; let $\psi_2$ be the prefix $\exists\, c_1, \ldots, \exists\, c_t$. Now $\langle \Phi\,\boxed{S_1}\,,\, \Phi\,\boxed{S_2}\,\rangle$ is reducible to $\langle \psi_1\,\boxed{S_1}\,,\, \psi_2\,\boxed{S_2}\,\rangle$ using operation $E$. Since $\psi_1, \psi_2$ contain no common parameters, then $(\psi_1, \psi_2)$ is obviously a special pair of prefixes. And $\langle\, \boxed{S_1}\,,\, \boxed{S_2}\,\rangle$ is reducible to $\langle \mathscr{D}_1, \mathscr{D}_2 \rangle$, so by Lemma I there is a *special* pair $(\psi_1', \psi_2')$ such that $\langle \psi_1\,\boxed{S_1}\,, \psi_2\,\boxed{S_2}\,\rangle$ is $A$, $B$, $C$, $D$ reducible to $\langle \psi_1'\mathscr{D}_1, \psi_2'\mathscr{D}_2 \rangle$. Hence $\langle \Phi\,\boxed{S_1}\,,\, \Phi\,\boxed{S_2}\,\rangle$ is $A$, $B$, $C$, $D$, $E$ reducible to $\langle \psi_1'\mathscr{D}_1, \psi_2'\mathscr{D}_2 \rangle$. Now let $\mathscr{C}_1$ be the result of first deleting all closed blocks from $\mathscr{D}_1$, then all non-atomic elements from the remaining blocks, and all atomic elements whose conjugates do not occur in any blocks of $\mathscr{D}_2$. Let $\sigma_1$ be the result of deleting all quantifiers $q\, a$ from $\psi_1'$ such that $a$ does not occur in $\mathscr{C}_1$. Similarly we define $\mathscr{C}_2$ and $\sigma_2$. Now $\langle \psi_1'\mathscr{D}_1, \psi_2'\mathscr{D}_2 \rangle$ is reducible to $\langle \sigma_1\mathscr{C}_1, \sigma_2\mathscr{C}_2 \rangle$ by operation $F$. It remains to show that $\langle \sigma_1\mathscr{C}_1, \sigma_2\mathscr{C}_2 \rangle$ is a basic pair.

Clearly $\mathscr{C}_1$, $\mathscr{C}_2$ satisfy conditions (1), (2), (3) of our definition of a basic pair; the point now is to verify that $\sigma_2$ is the dual of $\sigma_1$. Well, since $(\psi_1', \psi_2')$ is a special pair, so is $(\sigma_1, \sigma_2)$ (why?). Since $(\sigma_1, \sigma_2)$ is special, it remains only to show that $\sigma_1, \sigma_2$ contain exactly the same parameters (and this will imply that $\sigma_2 = \sigma_1^0$). Well suppose $k$ is a parameter of $\sigma_1$. Then $k$ is in $\mathscr{C}_1$, but $k$ is not in the set $\{b_1, \ldots, b_j\}$ (why?). The parameters of $\mathscr{C}_1$ are obviously the same as the parameters of $\mathscr{C}_2$, hence $k$ is in $\mathscr{C}_2$. Hence $k$ is in $\mathscr{D}_2$. But $k$ is not in the set $\{b_1, \ldots, b_j\}$, therefore $k$ must be in the prefix $\psi_2'$ (why?). Now the only parameters of $\psi_2'$ not in $\sigma_2$ are those of $\mathscr{D}_2$ not in $\mathscr{C}_2$. But $k$ is in $\mathscr{C}_2$, so $k$ is in $\sigma_2$. Thus all parameters of $\sigma_1$ are in $\sigma_2$. A similar argument shows that all parameters of $\sigma_2$ are in $\sigma_1$. This concludes the proof.

**Discussion.** Suppose $\langle \Phi\,\boxed{S_1}\,,\, \Phi\,\boxed{S_2}\,\rangle$ is $A - F$ reducible to a basic pair $\langle \sigma_1\mathscr{C}_1, \sigma_2\mathscr{C}_2 \rangle$; let $X_1$ be a translation of $\sigma_1\mathscr{C}_1$ and $X_2$ be a translation of $\sigma_2\mathscr{C}_2$. We assert that $X_1$ must be an interpolation formula for the stroke sequent $S_1 | S_2$ (and thus Theorem 3 affords yet another proof of Craig's interpolation lemma—rather closer in spirit to the original proof of Craig). For we know that $\mathscr{C}_1$, $\mathscr{C}_2$ contain the same predicates and parameters; $\sigma_1, \sigma_2$ contain the same parameters, hence the predicates of $X_1, X_2$ are the same, and the parameters of $X_1$ are those of $\mathscr{C}_1$ not in $\sigma_1$; the parameters of $X_2$ are those of $\mathscr{C}_2$ not in $\sigma_2$, hence $X_1$, $X_2$ contain the same parameters. All predicates and parameters of $X_1$ are in $S_1$; all predicates and parameters of $X_2$ are in $S_2$, thus all predicates and parameters of $X_1$ are in both $S_1$ and $S_2$. Finally, $S_1$ implies $X_1$, $S_2$ implies $X_2$, but $X_1$ is incompatible with $X_2$, so $S_2$ implies $\sim X_1$. Thus $X_1$ is an interpolation formula for $S_1 | S_2$.

## § 2. Linear Reasoning

Theorem 3 can be converted to a theorem on linear reasoning in the following manner.

By a *dual configuration* we shall mean an ordered pair $\langle \mathscr{C}, 0 \rangle$—also written $\mathscr{C}^0$—where $\mathscr{C}$ is a configuration. The "label" 0 is to be thought of as directions for interpreting the configuration differently—i.e. reading the comma inside the blocks as " $\vee$ " rather than " $\wedge$ ", and interpreting juxtaposition of blocks as " $\wedge$ " rather than " $\vee$ ". To be more precise, let $\mathscr{C}$ be a configuration $\boxed{B_1} \cdots \boxed{B_n}$. Then we say that $\mathscr{C}^0$ is *true* under a given interpretation $I$ if *every* block of $\mathscr{C}$ contains *at least* one element which is true under $I$. And we define a *translation* of $\mathscr{C}^0$ not as $\hat{B}_1 \vee \cdots \vee \hat{B}_n$ (which is a translation of $\mathscr{C}$), but rather as $\check{B}_1 \wedge \cdots \wedge \check{B}_n$. We shall sometimes write $\{B_1\} \cdots \{B_n\}$ for $(\boxed{B_1} \cdots \boxed{B_n})^0$.

Note that a one-block configuration $\boxed{X}$ whose block contains only one element is equivalent to the dual configuration $\{X\}$ (i.e. to $\boxed{X}^0$). Note also that if $\mathscr{C}$ is closed, the translation of $\mathscr{C}$ is truth-functionally unsatisfiable, whereas the translation of $\mathscr{C}^0$ is a tautology.

For any configuration $\mathscr{C}$, by its *conjugate* configuration $\bar{\mathscr{C}}$ we mean the result of replacing every element $X$ of every block by its conjugate $\bar{X}$. We note the following facts:

(a) Under any interpretation, $\bar{\mathscr{C}}$ has the opposite truth-value to $\mathscr{C}^0$.

(b) $\mathscr{C}_1$ implies $\mathscr{C}_2^0$ iff $\mathscr{C}_1$ is incompatible with $\bar{\mathscr{C}}_2$.

(c) $\sigma_1 \mathscr{C}_1^0$ implies $\sigma_2 \mathscr{C}_2^0$ iff $\sigma_2^0 \bar{\mathscr{C}}_2$ implies $\sigma_1^0 \bar{\mathscr{C}}_1$.

We shall collectively refer to operations $A, B, C, D, E, F$—as applied to single quantified configurations—as $L_1$-*operations*. Now we shall define the "duals" of $L_1$-operations; these are to be applied to quantified *dual* configurations. We shall say that $\sigma_2 \mathscr{C}_2^0$ is obtainable from $\sigma_1 \mathscr{C}_1^0$ by a *dual* $L_1$-operation iff $\sigma_1^0 \bar{\mathscr{C}}_1$ is obtainable from $\sigma_2^0 \bar{\mathscr{C}}_2$ by that $L_1$-operation. We shall refer to dual $L_1$-operations as $L_2$-*operations*. In detail, the $L_2$-operations are

$$A^0: \quad \frac{\sigma \mathscr{C}^0 \{S, \beta_1, \beta_2\}}{\sigma \mathscr{C}^0 \{S, \beta\}}$$

$$B^0: \quad \frac{\sigma \mathscr{C}^0 \{S, \alpha_1\} \{S, \alpha_2\}}{\sigma \mathscr{C}^0 \{S, \alpha\}}$$

$C^0$:   $\dfrac{\sigma \mathscr{C}^0 \{S, \delta, \delta(a)\}}{\sigma' \mathscr{C}^0 \{S, \delta\}}$ , where $\sigma'$ is defined as follows:

                 (i)   if $a$ does not occur in $\sigma$, then $\sigma' = \sigma$.

                 (ii)   if $a$ occurs in $\mathscr{C}^0 \{S, \delta\}$, then $\sigma' = \sigma$.

                 (iii)   if $qa$ occurs in $\sigma$ but $a$ does not occur in $\mathscr{C}^0 \{S, \delta\}$ then $\sigma'$ is the result of deleting $qa$ from $\sigma$.

$D^0$:   $\dfrac{\sigma \forall a \mathscr{C}^0 \{S, \gamma(a)\}}{\sigma \mathscr{C}^0 \{S, \gamma\}}$       $\dfrac{\sigma \mathscr{C}^0 \{S, \gamma(a)\}}{\sigma \mathscr{C}^0 \{S, \gamma\}}$

where $a$ does not    where $a$ does not occur in $\mathscr{C}^0 \{S, \gamma\}$ nor in
occur in $\mathscr{C}^0 \{S, \gamma\}$    $\gamma(a)$.

$E^0$:   (Universal Instantiation)

$$\frac{\sigma \forall a \mathscr{C}^0}{\sigma \mathscr{C}^0}$$

$F^0$:   $\dfrac{\sigma \mathscr{C}_1^0}{\sigma' \mathscr{C}_2^0}$ , where $\mathscr{C}_2$ is the result of adding a closed block to $\mathscr{C}_1$ or adding an arbitrary element to some block of $\mathscr{C}_1$ and $\sigma'$ is the result of taking those parameters $a_1, \ldots, a_n$ which occur in $\mathscr{C}_2$ but not in $\sigma$ and for each $a_i$, inserting either $\forall a_i$ or $\exists a_i$ anywhere in the prefix $\sigma$.

In addition to the $L_1$-operations and $L_2$-operations, we finally consider the operation

$S$:   (Switching Operation)

$\dfrac{\sigma \mathscr{C}_1}{\sigma \mathscr{C}_2^0}$ , where $\mathscr{C}_1$ strongly clashes with $\bar{\mathscr{C}}_2$.

We use the term "$L$-operations" collectively for $L_1, L_2$ and $S$ operations. We say that $\sigma_2 \mathscr{C}_2^0$ is $L$-deducible from $\sigma_1 \mathscr{C}_1$ if there exists a sequence of lines (called an $L$-deduction of $\sigma_2 \mathscr{C}_2^0$ from $\sigma_1 \mathscr{C}_1$), starting with $\sigma_1 \mathscr{C}_1$ and ending with $\sigma_2 \mathscr{C}_2^0$, such that each line (except the first) is obtainable from the preceding line by an application of an $L$-operation. (Every $L$-deduction must obviously begin with $L_1$-steps, then comes just one $S$-step, and then come $L_2$-steps.) In all $L$-operations, the premise implies the conclusion, hence if $\Phi \boxed{S_2}^\circ$ is $L$-deducible from $\Phi \boxed{S_1}$, then the sequent $S_1 \rightarrow S_2$ is valid. Theorem 3 now easily yields

**Theorem 4.** *(Completeness of Linear Reasoning). If $S_1 \rightarrow S_2$ is valid, then there exists an $L$-deduction of $\Phi \boxed{S_2}^0$ from $\Phi \boxed{S_1}$.*

**Proof.** Suppose $S_1 \to S_2$ is valid. Then $S_1$ is incompatible with $\bar{S}_2$. Then by Theorem 3, $\langle \Phi \boxed{S_1}, \Phi \boxed{\bar{S}_2} \rangle$ is $A - F$ reducible to a basic pair $\langle \sigma\mathscr{C}_1, \sigma^0\mathscr{C}_2 \rangle$. This means that $\sigma\mathscr{C}_1$ is $L_1$-deducible from $\Phi \boxed{S_1}$ (i.e. $L$-deducible using $L_1$-operations) and $\sigma^0\mathscr{C}_2$ is $L_1$-deducible from $\Phi \boxed{\bar{S}_2}$. Since $\mathscr{C}_1$ strongly clashes with $\mathscr{C}_2$, then $\sigma(\bar{\mathscr{C}}_2)^0$ is obtainable from $\sigma\mathscr{C}_1$ by an $S$-step. Since $\sigma^0\mathscr{C}_2$ is $L_1$-deducible from $\Phi \boxed{\bar{S}_2}$, then $\Phi \boxed{S_2}^0$ is $L_2$-deducible from $\sigma(\bar{\mathscr{C}}_2)^0$, which in turn is $S$-deducible from $\sigma\mathscr{C}_1$, which in turn is $L_1$-deducible from $\Phi \boxed{S_1}$. Thus $\Phi \boxed{S_2}^0$ is $L$-deducible from $\Phi \boxed{S_1}$.

**Remark.** Consider an $L$-deduction of $\Phi \boxed{S_2}^0$ from $\Phi \boxed{S_1}$; let $\sigma\mathscr{C}$ be the last configuration of the deduction (the next line is the first dual configuration). Then any translation of $\sigma\mathscr{C}$ is an interpolation formula for the sequent $S_1 \to S_2$ (cf. the discussion following the proof of Theorem 3).

The completeness of our system of linear reasoning derives ultimately from our completeness theorem for clashing tableaux. Two other new systems of linear reasoning—which arise naturally from our completeness theorem for clashing prenex tableaux and from the strong symmetric form of the fundamental theorem—are presented below.

## § 3. Linear Reasoning for Prenex Formulas

If we consider only prenex formulas, then we need to use only configurations (and dual configurations) with only one block. (This is a counterpart to the fact that prenex tableaux have only one branch.) We shall now identify a set $S$ (of prenex sentences) with the 1-block configuration $\boxed{S}$ (and by $S^0$ we shall mean the dual configuration $\boxed{S}^0$). Using the completeness theorem for $P$-clashing prenex tableaux, one easily proves

**Theorem 5.** *If $S_1$, $S_2$ are incompatible sets of prenex sentences, then $\langle S_1, S_2 \rangle$ is $C, D$ reducible (i.e. reducible using just operations $C, D$) to a pair $\langle S'_1, S'_2 \rangle$ such that $S'_1 \cup S'_2$ is truth-functionally unsatisfiable.*

Next it is easily verified that Lemma I (preceding Theorem 3) holds, reading "$C - D$ reducible" for "reducible". We again consider operation $E$, and in place of operation $F$, the following operation:

$F'$: $\dfrac{\sigma_1 S_1}{\sigma_2 S_2}$, provided $\hat{S}_1 \supset \hat{S}_2$ is a tautology, and all predicates and parameters of $S_2$ occur in $S_1$, and $\sigma_2$ results from $\sigma_1$ by deleting all quantifiers $qa$ such that $a$ does not occur in $S_2$.

The following is the analogue of Theorem 3 for prenex formulas:

**Theorem 6.** *If $S_1, S_2$ are incompatible sets of prenex sentences, then $\langle \Phi S_1, \Phi S_2 \rangle$ is C, D, E, F' reducible to a pair $\langle \sigma_1 S_1', \sigma_2 S_2' \rangle$ such that $S_1' \cup S_2'$ is truth-functionally unsatisfiable, and $S_1', S_2'$ contain the same predicates and parameters, and $\sigma_2$ is the dual of $\sigma_1$.*

Theorem 6 is proved much in the manner of Theorem 3, using Theorem 5 in place of Theorem 2, and also using Craig's Interpolation Lemma for *propositional* logic (verify!).

Next we consider in place of operation $S$, the following operation:

$S'$: $\dfrac{\sigma S_1}{\sigma S_2^0}$, where $\hat{S}_1 \supset \hat{S}_2$ is a tautology.

By the *dual $F'^0$* of operation $F'$ we mean:

$\dfrac{\sigma_1 S_1^0}{\sigma_2 S_2^0}$, provided $\hat{S}_1 \supset \hat{S}_2$ is a tautology, and all predicates and parameters of $S_1$ are in $S_2$, and all parameters in $\sigma_2$ occur in $S_2^0$, and $\sigma_1$ is the result of deleting from $\sigma_2$ all quantifiers $q\,a$ such that $a$ does not occur in $S_1$.

Now Theorem 6 easily yields

**Theorem 7.** *(A Completeness Theorem for Linear Reasoning for Prenex Formulas). If $S_1 \to S_2$ is valid, then $\Phi S_2^0$ is deducible from $\Phi S_1$ using operations C, D, E, F' and their duals, and operation S'.*

The reader might find it of interest to compare this system with the original system of Craig [1].

## § 4. A System Based on the Strong Symmetric Form of the Fundamental Theorem

This system (perhaps the simplest of all) again uses 1-block configurations but does not appeal to prenex normal form. It is like the preceding system except that we replace operations $C, D$ by the following operations:

$C'$: $\dfrac{\sigma \boxed{S}}{\sigma' \boxed{S, \gamma \supset \gamma(a)}}$, where $\gamma$ is a Boolean descendant of some element of $S$, and $\sigma'$ is $\sigma$ if $a$ is in $\sigma$ or $S$ or not in $\gamma(a)$; otherwise $\sigma'$ is the result of putting $\exists\,a$ or $\forall\,a$ anywhere in $\sigma$.

$D'$: $\dfrac{\sigma \boxed{S}}{\sigma' \boxed{S, \delta \supset \delta(a)}}$, where $\delta$ is a Boolean descendant of some element of $S$, $a$ does not occur in $S$ nor in $\delta$, and $\sigma'$ is $\sigma$ or $\sigma \,\exists\, a$ depending respectively upon whether $a$ does or does not occur in $\delta(a)$.

The definition of the dual operations $C'^0, D'^0$ should be obvious.

**Theorem 8.** *For any sets $S_1, S_2$ of sentences (not necessarily prenex), if $S_1 \rightarrow S_2$ is valid, then $S_2^0$ is deducible from $S_1$ using operations $C', D', E, F'$ and their duals and the operation $S'$.*

The proof of Theorem 8 should be easily obtainable by the reader—it uses in place of Theorem 3 the strong symmetric form of the Fundamental Theorem (or more directly, the completeness of the system $QQ$ in the Gentzen type formulation (cf. end of preceding chapter)) as well as Craig's Interpolation Lemma for propositional logic.

# References

Anderson, A. R., and Nuel D. Belnap, Jr.: [1] A simple proof of Gödel's Completeness theorem (abstract). J. S. L. **24, 4,** 320 (1959).

Beth, E. W.: [1] The Foundations of Mathematics. North Holland 1959.

Church, A.: [1] Introduction to Mathematical Logic I. Princeton 1956.

Craig, W.: [1] Linear Reasoning. A New Form of the Herbrand-Gentzen theorem. J. S. L. **22,** 250–268 (1957).

Gentzen, G.: [1] Untersuchungen über das logische Schließen. Mathematische Zeitschrift. **39,** 176–210, 405–431 (1935).

Hintikka, K. J. J.: [1] Form and content in quantification theory. Acta Philosophica Fennica. **8,** 7–55 (1955).

Kleene, S. C.: [1] Introduction to Metamathematics. Princeton: Van-Nostrand 1952.

Lis, Z.: [1] Wynikanie semantyczne a wynikanie formalne ("Logical consequence, semantic and formal"). Studia Logica **10,** 39–60 (1960).

Lyndon, R. C.: [1] Notes on Logic. Princeton: Van-Nostrand 1966.

Mendelson, E.: [1] Introduction to Mathematical Logic. Princeton: Van-Nostrand 1964.

Patton, T.: [1] A system of quantificational deduction. Notre Dame Journal of Formal Logic. **IV, 3,** (1963).

Quine: [1] Methods of Logic. New York: Henry Holt and Co. 1959.

Smullyan, R. M.: [1] Trees and Nest Structures. J. S. L. **31,** 303–321 (1966).

— [2] A Unifying Principle in Quantification Theory. Proceedings of the National Academy of Sciences, June 1963.

— [3] Analytic natural deduction. J. S. L. **30,** 123–139 (1965).

— [4] Abstract Quantification Theory. Proc. Conference on Intuitionism and Proof Theory. North Holland 1969.

# Subject Index

# A CATALOG OF SELECTED
# DOVER BOOKS
## IN SCIENCE AND MATHEMATICS

# A CATALOG OF SELECTED
# DOVER BOOKS
## IN SCIENCE AND MATHEMATICS

QUALITATIVE THEORY OF DIFFERENTIAL EQUATIONS, V.V. Nemytskii and V.V. Stepanov. Classic graduate-level text by two prominent Soviet mathematicians covers classical differential equations as well as topological dynamics and ergodic theory. Bibliographies. 523pp. 5⅜ × 8½.　　　65954-2 Pa. $10.95

MATRICES AND LINEAR ALGEBRA, Hans Schneider and George Phillip Barker. Basic textbook covers theory of matrices and its applications to systems of linear equations and related topics such as determinants, eigenvalues and differential equations. Numerous exercises. 432pp. 5⅜ × 8½.　　　66014-1 Pa. $10.95

QUANTUM THEORY, David Bohm. This advanced undergraduate-level text presents the quantum theory in terms of qualitative and imaginative concepts, followed by specific applications worked out in mathematical detail. Preface. Index. 655pp. 5⅜ × 8½.　　　65969-0 Pa. $13.95

ATOMIC PHYSICS (8th edition), Max Born. Nobel laureate's lucid treatment of kinetic theory of gases, elementary particles, nuclear atom, wave-corpuscles, atomic structure and spectral lines, much more. Over 40 appendices, bibliography. 495pp. 5⅜ × 8½.　　　65984-4 Pa. $12.95

ELECTRONIC STRUCTURE AND THE PROPERTIES OF SOLIDS: The Physics of the Chemical Bond, Walter A. Harrison. Innovative text offers basic understanding of the electronic structure of covalent and ionic solids, simple metals, transition metals and their compounds. Problems. 1980 edition. 582pp. 6⅛ × 9¼.　　　66021-4 Pa. $15.95

BOUNDARY VALUE PROBLEMS OF HEAT CONDUCTION, M. Necati Özisik. Systematic, comprehensive treatment of modern mathematical methods of solving problems in heat conduction and diffusion. Numerous examples and problems. Selected references. Appendices. 505pp. 5⅜ × 8½.　　　65990-9 Pa. $12.95

A SHORT HISTORY OF CHEMISTRY (3rd edition), J.R. Partington. Classic exposition explores origins of chemistry, alchemy, early medical chemistry, nature of atmosphere, theory of valency, laws and structure of atomic theory, much more. 428pp. 5⅜ × 8½. (Available in U.S. only)　　　65977-1 Pa. $10.95

A HISTORY OF ASTRONOMY, A. Pannekoek. Well-balanced, carefully reasoned study covers such topics as Ptolemaic theory, work of Copernicus, Kepler, Newton, Eddington's work on stars, much more. Illustrated. References. 521pp. 5⅜ × 8½.　　　65994-1 Pa. $12.95

PRINCIPLES OF METEOROLOGICAL ANALYSIS, Walter J. Saucier. Highly respected, abundantly illustrated classic reviews atmospheric variables, hydrostatics, static stability, various analyses (scalar, cross-section, isobaric, isentropic, more). For intermediate meteorology students. 454pp. 6⅛ × 9¼. 65979-8 Pa. $14.95

GEOMETRY OF COMPLEX NUMBERS, Hans Schwerdtfeger. Illuminating, widely praised book on analytic geometry of circles, the Moebius transformation, and two-dimensional non-Euclidean geometries. 200pp. 5⅜ × 8¼.
63830-8 Pa. $8.95

MECHANICS, J.P. Den Hartog. A classic introductory text or refresher. Hundreds of applications and design problems illuminate fundamentals of trusses, loaded beams and cables, etc. 334 answered problems. 462pp. 5⅜ × 8½.   60754-2 Pa. $9.95

TOPOLOGY, John G. Hocking and Gail S. Young. Superb one-year course in classical topology. Topological spaces and functions, point-set topology, much more. Examples and problems. Bibliography. Index. 384pp. 5⅜ × 8¼.
65676-4 Pa. $9.95

STRENGTH OF MATERIALS, J.P. Den Hartog. Full, clear treatment of basic material (tension, torsion, bending, etc.) plus advanced material on engineering methods, applications. 350 answered problems. 323pp. 5⅜ × 8½.   60755-0 Pa. $8.95

ELEMENTARY CONCEPTS OF TOPOLOGY, Paul Alexandroff. Elegant, intuitive approach to topology from set-theoretic topology to Betti groups; how concepts of topology are useful in math and physics. 25 figures. 57pp. 5⅜ × 8½.
60747-X Pa. $3.50

ADVANCED STRENGTH OF MATERIALS, J.P. Den Hartog. Superbly written advanced text covers torsion, rotating disks, membrane stresses in shells, much more. Many problems and answers. 388pp. 5⅜ × 8½.   65407-9 Pa. $9.95

COMPUTABILITY AND UNSOLVABILITY, Martin Davis. Classic graduate-level introduction to theory of computability, usually referred to as theory of recurrent functions. New preface and appendix. 288pp. 5⅜ × 8½. 61471-9 Pa. $7.95

GENERAL CHEMISTRY, Linus Pauling. Revised 3rd edition of classic first-year text by Nobel laureate. Atomic and molecular structure, quantum mechanics, statistical mechanics, thermodynamics correlated with descriptive chemistry. Problems. 992pp. 5⅜ × 8½.   65622-5 Pa. $19.95

AN INTRODUCTION TO MATRICES, SETS AND GROUPS FOR SCIENCE STUDENTS, G. Stephenson. Concise, readable text introduces sets, groups, and most importantly, matrices to undergraduate students of physics, chemistry, and engineering. Problems. 164pp. 5⅜ × 8½.   65077-4 Pa. $6.95

THE HISTORICAL BACKGROUND OF CHEMISTRY, Henry M. Leicester. Evolution of ideas, not individual biography. Concentrates on formulation of a coherent set of chemical laws. 260pp. 5⅜ × 8½.   61053-5 Pa. $6.95

THE PHILOSOPHY OF MATHEMATICS: An Introductory Essay, Stephan Körner. Surveys the views of Plato, Aristotle, Leibniz & Kant concerning propositions and theories of applied and pure mathematics. Introduction. Two appendices. Index. 198pp. 5⅜ × 8½.   25048-2 Pa. $7.95

THE DEVELOPMENT OF MODERN CHEMISTRY, Aaron J. Ihde. Authoritative history of chemistry from ancient Greek theory to 20th-century innovation. Covers major chemists and their discoveries. 209 illustrations. 14 tables. Bibliographies. Indices. Appendices. 851pp. 5⅜ × 8½.   64235-6 Pa. $18.95

DE RE METALLICA, Georgius Agricola. The famous Hoover translation of greatest treatise on technological chemistry, engineering, geology, mining of early modern times (1556). All 289 original woodcuts. 638pp. 6¾ × 11.
60006-8 Pa. $18.95

SOME THEORY OF SAMPLING, William Edwards Deming. Analysis of the problems, theory and design of sampling techniques for social scientists, industrial managers and others who find statistics increasingly important in their work. 61 tables. 90 figures. xvii + 602pp. 5⅜ × 8½.
64684-X Pa. $15.95

THE VARIOUS AND INGENIOUS MACHINES OF AGOSTINO RAMELLI: A Classic Sixteenth-Century Illustrated Treatise on Technology, Agostino Ramelli. One of the most widely known and copied works on machinery in the 16th century. 194 detailed plates of water pumps, grain mills, cranes, more. 608pp. 9 × 12.
28180-9 Pa. $24.95

LINEAR PROGRAMMING AND ECONOMIC ANALYSIS, Robert Dorfman, Paul A. Samuelson and Robert M. Solow. First comprehensive treatment of linear programming in standard economic analysis. Game theory, modern welfare economics, Leontief input-output, more. 525pp. 5⅜ × 8½.
65491-5 Pa. $14.95

ELEMENTARY DECISION THEORY, Herman Chernoff and Lincoln E. Moses. Clear introduction to statistics and statistical theory covers data processing, probability and random variables, testing hypotheses, much more. Exercises. 364pp. 5⅜ × 8½.
65218-1 Pa. $9.95

THE COMPLEAT STRATEGYST: Being a Primer on the Theory of Games of Strategy, J.D. Williams. Highly entertaining classic describes, with many illustrated examples, how to select best strategies in conflict situations. Prefaces. Appendices. 268pp. 5⅜ × 8½.
25101-2 Pa. $7.95

MATHEMATICAL METHODS OF OPERATIONS RESEARCH, Thomas L. Saaty. Classic graduate-level text covers historical background, classical methods of forming models, optimization, game theory, probability, queueing theory, much more. Exercises. Bibliography. 448pp. 5⅜ × 8¼.
65703-5 Pa. $12.95

CONSTRUCTIONS AND COMBINATORIAL PROBLEMS IN DESIGN OF EXPERIMENTS, Damaraju Raghavarao. In-depth reference work examines orthogonal Latin squares, incomplete block designs, tactical configuration, partial geometry, much more. Abundant explanations, examples. 416pp. 5⅜ × 8¼.
65685-3 Pa. $10.95

THE ABSOLUTE DIFFERENTIAL CALCULUS (CALCULUS OF TENSORS), Tullio Levi-Civita. Great 20th-century mathematician's classic work on material necessary for mathematical grasp of theory of relativity. 452pp. 5⅜ × 8½.
63401-9 Pa. $9.95

VECTOR AND TENSOR ANALYSIS WITH APPLICATIONS, A.I. Borisenko and I.E. Tarapov. Concise introduction. Worked-out problems, solutions, exercises. 257pp. 5⅜ × 8¼.
63833-2 Pa. $7.95

**CHALLENGING MATHEMATICAL PROBLEMS WITH ELEMENTARY SOLUTIONS,** A.M. Yaglom and I.M. Yaglom. Over 170 challenging problems on probability theory, combinatorial analysis, points and lines, topology, convex polygons, many other topics. Solutions. Total of 445pp. 5⅜ × 8½. Two-vol. set.

Vol. I 65536-9 Pa. $7.95
Vol. II 65537-7 Pa. $6.95

**FIFTY CHALLENGING PROBLEMS IN PROBABILITY WITH SOLUTIONS,** Frederick Mosteller. Remarkable puzzlers, graded in difficulty, illustrate elementary and advanced aspects of probability. Detailed solutions. 88pp. 5⅜ × 8½.

65355-2 Pa. $4.95

**EXPERIMENTS IN TOPOLOGY,** Stephen Barr. Classic, lively explanation of one of the byways of mathematics. Klein bottles, Moebius strips, projective planes, map coloring, problem of the Koenigsberg bridges, much more, described with clarity and wit. 43 figures. 210pp. 5⅜ × 8½. 25933-1 Pa. $5.95

**RELATIVITY IN ILLUSTRATIONS,** Jacob T. Schwartz. Clear nontechnical treatment makes relativity more accessible than ever before. Over 60 drawings illustrate concepts more clearly than text alone. Only high school geometry needed. Bibliography. 128pp. 6⅛ × 9¼. 25965-X Pa. $6.95

**AN INTRODUCTION TO ORDINARY DIFFERENTIAL EQUATIONS,** Earl A. Coddington. A thorough and systematic first course in elementary differential equations for undergraduates in mathematics and science, with many exercises and problems (with answers). Index. 304pp. 5⅜ × 8½. 65942-9 Pa. $8.95

**FOURIER SERIES AND ORTHOGONAL FUNCTIONS,** Harry F. Davis. An incisive text combining theory and practical example to introduce Fourier series, orthogonal functions and applications of the Fourier method to boundary-value problems. 570 exercises. Answers and notes. 416pp. 5⅜ × 8½. 65973-9 Pa. $9.95

**THE THEORY OF BRANCHING PROCESSES,** Theodore E. Harris. First systematic, comprehensive treatment of branching (i.e. multiplicative) processes and their applications. Galton-Watson model, Markov branching processes, electron-photon cascade, many other topics. Rigorous proofs. Bibliography. 240pp. 5⅜ × 8½. 65952-6 Pa. $6.95

**AN INTRODUCTION TO ALGEBRAIC STRUCTURES,** Joseph Landin. Superb self-contained text covers "abstract algebra": sets and numbers, theory of groups, theory of rings, much more. Numerous well-chosen examples, exercises. 247pp. 5⅜ × 8½. 65940-2 Pa. $7.95

---

*Prices subject to change without notice.*
Available at your book dealer or write for free Mathematics and Science Catalog to Dept. GI, Dover Publications, Inc., 31 East 2nd St., Mineola, N.Y. 11501. Dover publishes more than 175 books each year on science, elementary and advanced mathematics, biology, music, art, literature, history, social sciences and other areas.